Praise for *Borrowed Tim*

"Powerful . . . Agonizing . . . Rises to shattering eloquence. The category of gay literature no longer applies: we enter the universal arena of human loss." —*Newsweek*

"Enormously affecting . . . impossible to put down . . . The sense of loss contained in this book transcends one man, one gay couple, the gay community . . . Monette strikes a universal chord."
—*The Winston-Salem Journal*

"*Borrowed Time* brings the plague years home as no other book does. It is impossible to read this love story without weeping . . . The effect is so over-powering, so emotion-charged that at times we simply have to stop reading." —*Newsday*

"Intensely felt and finely written . . . Likely to survive the tragic period it records . . . Completely human and completely real . . . One experience we cannot forget." —*The Philadelphia Inquirer*

"Intense . . . Heartbreaking . . . A necessary document for a time when many have been forced to negotiate their contracts with death." —*Village Voice*

"This is a riveting story—the dark, more personal side of Randy Shilts's *And the Band Played On.*" —*Kirkus Reviews*

Winner of the
PEN West Literary Award for Best Nonfiction

Nominated for the
National Book Critic's Circle Award

# BORROWED TIME

## ALSO BY PAUL MONETTE

### POEMS

The Carpenter at the Asylum

No Witnesses

Love Alone: 18 Elegies for Rog

West of Yesterday, East of Summer:
New and Selected Poems (1973–1993)

### NOVELS

Taking Care of Mrs. Carroll

The Gold Diggers

The Long Shot

Lightfall

Afterlife

Halfway Home

### NONFICTION

Becoming a Man: Half a Life Story

Last Watch of the Night

# BORROWED TIME

## AN AIDS MEMOIR

PAUL MONETTE

A HARVEST BOOK • HARCOURT, INC.

ORLANDO AUSTIN NEW YORK
SAN DIEGO *TORONTO* LONDON

Library of Congress-in-Publication Data
Monette, Paul
Borrowed time.
1. Monette, Paul—Health.
2. AIDS (Disease)—Patients—United States—Biography.
I. Title.
RC607.A26M66 1988 362.1'96792'0922 [B] 88-7215
ISBN 0-15-113598-3
ISBN 0-15-600581-6 (PB)
ISBN 978-0-15-600581-4 (PB)

Text set in Sabon
Designed by Gracie Artemis
Printed in the United States of America
First Harvest Edition 1998
DOC 10 9 8 7

TO DENNIS COPE

*Unsung the noblest deed will die*

—Pindar, Fragment 120

# BORROWED TIME

# ·1·

I don't know if I will live to finish this. Doubtless there's a streak of self-importance in such an assertion, but who's counting? Maybe it's just that I've watched too many sicken in a month and die by Christmas, so that a fatal sort of realism comforts me more than magic. All I know is this: The virus ticks in me. And it doesn't care a whit about our categories—when is full-blown, what's AIDS-related, what is just sick and tired? No one has solved the puzzle of its timing. I take my drug from Tijuana twice a day. The very friends who tell me how vigorous I look, how well I seem, are the first to assure me of the imminent medical breakthrough. What they don't seem to understand is, I used up all my optimism keeping my friend alive. Now that he's gone, the cup of my own health is neither half full nor half empty. Just half.

Equally difficult, of course, is knowing where to start. The world around me is defined now by its endings and its closures—the date on the grave that follows the hyphen. Roger Horwitz, my beloved friend, died of complications of AIDS on October 22, 1986, nineteen months and ten days after his diagnosis. That is the only real date anymore, casting its ice shadow over all the secular holidays lovers mark their calendars by. Until that long night in October, it didn't seem possible that any day could supplant the brute equinox of March 12—the day of Roger's diagnosis in 1985, the day we began to live on the moon.

The fact is, no one knows where to start with AIDS. Now, in the seventh year of the calamity, my friends in L.A. can hardly recall what it felt like any longer, the time before the sickness. Yet we all watched the toll mount in New York, then in San Francisco, for years before it ever touched us here. It comes like a slowly dawning horror. At first you are equipped with a hundred different amulets to keep it far away. Then someone you know goes into the hospital, and suddenly you are at high noon in full battle gear. They have neglected to tell you that you will be issued no weapons of any sort. So you cobble together a weapon out of anything that lies at hand, like a prisoner honing a spoon handle into a stiletto. You fight tough, you fight dirty, but you cannot fight dirtier than it.

I remember a Saturday in February 1982, driving Route 10 to Palm Springs with Roger to visit his parents for the weekend. While Roger drove, I read aloud an article from *The Advocate*: "Is Sex Making Us Sick?" There was the slightest edge of irony in the query, an urban cool that seems almost bucolic now in its innocence. But the article didn't mince words. It was the first in-depth reporting I'd read that laid out the shadowy nonfacts of what till then had been the most fragmented of rumors. The first cases were reported to the Centers for Disease Control (CDC) only six months before, but they weren't in the newspapers, not in L.A. I note in my diary in December '81 ambiguous reports of a "gay cancer," but I know I didn't have the slightest picture of the thing. Cancer of the *what*? I would have asked, if anyone had known anything.

I remember exactly what was going through my mind while I was reading, though I can't now recall the details of the piece. I was thinking: How is this not me? Trying to find a pattern I was exempt from. It was a brand of denial I would watch grow exponentially during the next few years, but at the time I was simply relieved. Because the article appeared to be saying that there was a grim progression toward this undefined catastrophe, a set of preconditions—chronic hepatitis, repeated bouts of syphilis, exotic parasites. No wonder my first baseline response was to feel safe. It was *them*—by which I meant the fast-lane Fire Island crowd, the Sutro Baths, the world of High Eros.

Not us.

I grabbed for that relief because we'd been through a rough patch the previous autumn. Till then Roger had always enjoyed a sort of no-nonsense good health: not an abuser of anything, with a constitutional aversion to hypochondria, and not wed to his mirror save for a minor alarm as to the growing dimensions of his bald spot. In the seven years we'd been together I scarcely remember him having a cold or taking an aspirin. Yet in October '81 he had struggled with a peculiar bout of intestinal flu. Nothing special showed up in any of the blood tests, but over a period of weeks he experienced persistent symptoms that didn't neatly connect: pains in his legs, diarrhea, general malaise. I hadn't been feeling notably bad myself, but on the other hand I was a textbook hypochondriac, and I figured if Rog was harboring some kind of bug, so was I.

The two of us finally went to a gay doctor in the Valley for a further set of blood tests. It's a curious phenomenon among gay middle-class men that anything faintly venereal had better be taken to a doctor who's "on the bus." Is it a sense of fellow feeling perhaps, or a way of avoiding embarrassment? Do we really believe that only a doctor who's *our* kind can heal us of the afflictions that attach somehow to our secret hearts? There is so much magic to medicine. Of course we didn't know then that those few physicians with a large gay clientele were about to be swamped beyond all capacity to cope.

The tests came back positive for amoebiasis. Roger and I began the highly toxic treatment to kill the amoeba, involving two separate drugs and what seems in memory thirty pills a day for six weeks, till the middle of January. It was the first time I'd ever experienced the phenomenon of the cure making you sicker. By the end of treatment we were both weak and had lost weight, and for a couple of months afterward were susceptible to colds and minor infections.

It was only after the treatment was over that a friend of ours, diagnosed with amoebas by the same doctor, took his slide to the lab at UCLA for a second opinion. And that was my first encounter with lab error. The doctor at UCLA explained that the slide had been misread; the squiggles that looked like amoebas were in fact benign. The doctor shook his head and grumbled about "these guys who do their own lab work." Roger then retrieved his slide, took it over to UCLA and was told the same: no amoebas. We had just spent six weeks methodically ingesting poison for no reason at all.

So it wasn't the *Advocate* story that sent up the red flag for us. We'd been shaken by the amoeba business, and from that point on we operated at a new level of sexual caution. What is now called safe sex did not use to be so clearly defined. The concept didn't exist. But it was quickly becoming apparent, even then, that we couldn't wait for somebody else to define the parameters. Thus every gay man I know has had to come to a point of personal definition by way of avoiding the chaos of sexually transmitted diseases, or STD as we call them in the trade. There was obviously no one moment of conscious decision, a bolt of clarity on the shimmering freeway west of San Bernardino, but I think of that day when I think of the sea change. The party was going to have to stop. The evidence was too ominous: *We were making ourselves sick.*

Not that Roger and I were the life of the party. Roger especially didn't march to the different drum of *so many men, so little time*, the motto and anthem of the sunstruck summers of the mid-to-late seventies. He'd managed not to carry away from his adolescence the mark of too much repression, or indeed the yearning to make up for lost time. In ten years he had perhaps half a dozen contacts outside the main frame of our relationship, mostly when he was out of town on business. He was comfortable with relative monogamy, even at a time when certain quarters of the gay world found the whole idea trivial and bourgeois. I realize that in the world of the heterosexual there is a generalized lip service paid to exclusive monogamy, a notion most vividly honored in the breach. I leave the matter of morality to those with the gift of tongues; it was difficult enough for us to fashion a sexual ethics just for us. In any case, I was the one in the relationship who suffered from lost time. I was the one who would go after a sexual encounter as if it were an ice cream cone—casual, quick, good-bye.

But as I say, who's counting? I only want to make it plain to start with that we got very alert and very careful as far back as the winter of '82. That gut need for safety took hold and lingered, even as we got better again and strong. Thus I'm not entirely sure what I thought on another afternoon a year and a half later, when a friend of ours back from New York reported a conversation he'd had with a research man from Sloan-Kettering.

"He thinks all it takes is one exposure," Charlie said, this after months of articles about the significance of repeated exposure. More tenaciously than ever, we all wanted to believe the whole deepening tragedy was centered on those at the sexual frontiers who were fucking their brains out. The rest of us were fashioning our own little Puritan forts, as we struggled to convince ourselves that a clean slate would hold the nightmare at bay.

Yet with caution as our watchword starting in February of '82, Roger was diagnosed with AIDS three years later. So the turning over of new leaves was not to be on everybody's side. A lot of us were already ticking and didn't even know. The magic circle my generation is trying to stay within the borders of is only as real as the random past. Perhaps the young can live in the magic circle, but only if those of us who are ticking will tell our story. Otherwise it goes on being *us* and *them* forever, built like a wall higher and higher, till you no longer think to wonder if you are walling it out or in.

For us the knowing began in earnest on the first of September, 1983. I'd had a call a couple of days before from my closest friend, Cesar Albini, who'd just returned to San Francisco after a summer touring Europe with a group of students. He said he'd been having trouble walking because of a swollen gland in his groin, and he was going to the hospital to have it biopsied. He reassured me he was feeling fine and wasn't expecting anything ominous, but figured he'd check it out before school started again. AIDS didn't even cross my mind, though cancer did. Half joking, Cesar wondered aloud if he dared disturb our happy friendship with bad news.

"If it's bad," I said, "we'll handle it, okay?"

But I really didn't clutch with fear, or it was only a brief stab of the hypochondriacal sort. Roger and I were busy getting ready for a four-day trip to Big Sur, something we'd done almost yearly since moving to California in 1977. We were putting the blizzard of daily life on hold, looking forward to a dose of raw sublime that coincided with our anniversary—September 3, the day we met.

Cesar was forty-three, only ten months older than Roger. Born in Uruguay, possessed of a great heart and inexhaustible energy, he had studied in Europe and traveled all over, once spending four months going overland from Paris to China at a total cost of five hundred dollars. He was the first Uruguayan ever to enter Afghanistan through the mountains—on a camel, if I remember right. He spoke French, Italian, Spanish and English with equal fluency, and he tended to be the whole language department of a school. We'd both been teaching at secondary schools in Massachusetts when we met, and we goaded one another to make the move west that had always been our shared dream. Thus Cesar had relocated to San Francisco in July of '76, and Roger and I landed in L.A. four days after Thanksgiving the following year.

Cesar wasn't lucky in matters of the heart. He was still in the closet during his years back east, and the move to San Francisco was an extraordinary rite of passage for him. He always wanted a great love, but the couple of relationships he'd been involved in scarcely left the station. Still, he was very proud and indulged in no self-pity. He learned to accept the limited terms of the once-a-week relations he found in San Francisco, and broke through to the freedom of his own manhood without the mythic partner. The open sexual exultation that marked San Francisco in those days was something he rejoiced in.

Yet even though he went to the baths a couple of times a week, Cesar wasn't into anything *weird*—or that's how I might have put it at that stage of my own denial. No hepatitis, no history of VD, built tall and fierce—of course he was safe. The profile of AIDS continued to be mostly a matter of shadows. The L.A. *Times* wasn't covering it, though by then I had come to learn how embattled things had grown in New York. The Gay Men's Health Crisis was up to its ears in clients; Larry Kramer was screaming at the mayor; and the body count was appearing weekly in the *Native*. A writer I knew slightly was walking around with Kaposi's sarcoma. A young composer kept getting sicker and sicker, though he stubbornly didn't fit the CDC's hopelessly narrow categories, so that case was still officially a toss-up. And again, we're talking New York.

I came home at six on the evening of the first, and Roger met me gravely at the door. "There's a message from Cesar," he said. "It's not good."

Numbly I played back the answering machine, where so much appalling misery would be left on tape over the years to come, as if a record were crying out to be kept. "I have a little bit of bad news." Cesar's voice sounded strained, almost embarrassed. He left no details. I called and called him throughout the evening, convinced I was about to hear cancer news. The lymph nodes, of course—a hypochondriac knows all there is to know about the sites of malignancy. Already I was figuring what the treatments might be; no question in my mind but that it was treatable. I had Cesar practically cured by the time I reached Tom, a friend and former student of his. But as usual with me in crisis, I was jabbering and wouldn't let Tom get a word in. Finally he broke through: "He's got it."

"Got what?"

It's not till you first hear it attached to someone you love that you realize how little you know about it. My mind went utterly blank. The carefully constructed wall collapsed as if a 7.5 quake had rumbled under it. At that point I didn't even know the difference between KS and the opportunistic infections. I kept picturing that swollen gland in his groin, thinking: What's *that* got to do with AIDS? And a parallel track in my mind began careening with another thought: the swollen glands in my own groin, always dismissed by my straight doctor as herpes-related and "not a significant sign."

"We're not going to die young," Cesar used to say with a wag of his finger, his black Latin eyes dancing. "We won't get out of it *that* easily!" Then he would laugh and clap his hands, downing the coffee he always took with cream and four sugars. It looked like pudding.

I reached him very late that night and mouthed again the same words I'd said so bravely two days before: We'll deal with it. There is no end to the litany of reassurance that springs to your lips to ward away the specter. They've caught it early; you're fine; there's got to be some kind of treatment. That old chestnut, the imminent breakthrough. You fling these phrases instinctively, like pennies down a well. Cesar and I bent backward to calm each other. It was just a couple of lesions in the groin; you could hardly see them. And the reason everything was going to be all right was really very simple: We would fight this thing like demons.

But the hollowness and disbelief pursued Roger and me all the way up the gold coast. Big Sur was towering and bracing as ever—exalted as Homer's Ithaca, as Robinson Jeffers described it. We were staying at Ventana, the lavish inn high in the hills above the canyon of the Big Sur River. We used the inn as a base camp for our day-long hikes, returning in the evening to posh amenities worthy of an Edwardian big-game hunt. On the second morning we walked out to Andrew Molera Beach, where the Big Sur empties into the Pacific. Molera stretches unblemished for five miles down the coast, curving like a crescent moon, with weathered headlands clean as Scotland. It was a kind of holy place for Roger and me, like the yearly end of a quest.

"What if we got it?" I said, staring out at the otters belly up in the kelp beds, taking the sun.

I don't remember how we answered that, because of course there wasn't any answer. Merely to pose the question was by way of another shot at magic. Mention the unmentionable and it will go away, like shining a light around a child's bedroom to shoo the monster. The great ache we were feeling at that moment was for our stricken friend, and we were too ignorant still to envision the medieval tortures that might await him.

But I know that the roll of pictures I took that day was my first conscious memorializing of Roger and me, as if I could hold the present as security on the future. There's one of me on the beach, then a mirror image of him as we traded off the camera, both of us squinting in the clear autumn light with the river mouth behind. Back at the inn, I took a picture of Rog in a rope hammock, his blue eyes resting on me as if the camera weren't even there, in total equilibrium, nine years to the day since our paths crossed on Revere Street. His lips are barely curved in a quarter-smile, his hands at rest in his lap as the last wave of the westering sun washes his left side through the diamond weave of the rope.

How do I speak of the person who was my life's best reason? The most completely unpretentious man I ever met, modest and decent to such a degree that he seemed to release what was most real in everyone he knew. It was always a relief to be with Roger, not to have to play any games at all. By a safe mile he was the least flashy of all our bright circle of friends, but he spoke about books and the wide world he had journeyed with huge conviction and a hunger to know everything.

He had a contagious, impish sense of humor, especially about the folly of things, especially self-importance. Yet he was blissfully unfrivolous, without a clue as to what was "in." He had thought life through somehow and come out the other side with a proper respect for small pleasures. "*Quelle bonne soirée*," as Madeleine once exclaimed after dinner one night with us in Les Halles, a bistro called Pig's Foot. Wonderful evenings were second nature to us by then, with long walks at the end, especially when we traveled. Days we spent cavorting through museums, drunk on old things, like ten-year-olds loose in a castle. Roger loved nothing better than a one-on-one talk with a friend, and he had never lost track of a single one, all the way back to high school. The luck of the draw was mine, for I was the best and the most.

We met on the eve of Labor Day in 1974, at a dinner party at a mutual friend's apartment on Beacon Hill, just two days before Roger was to start work as an attorney at a stately firm in Boston. He was thirty-two; I was twenty-eight. Summer has always been good to me, even the bittersweet end, with the slant of yellow light, and I for one was in love before the night was done. I suppose we'd been waiting for each other all our lives. The business of coming out had been difficult for both of us, partly because of the closet nature of all relations in a Puritan town like Boston, partly because we were both so sure of what we wanted and it kept not coming to life.

"Spain!" Roger writes in his diary in 1959, after three days' hitchhike from Paris to Madrid. "If only I had a friend!"

For if there was no man out there who was equal and simpatico, then what was the point of being gay? The baggage and the shit you had to take were bad enough. But it all jogged into place when we met, everything I'd brooded over from the ancient Greeks to Whitman. It all ceased to be literary. My life was a sort of amnesia till then, longing for something that couldn't be true until I'd found the rest of me. Is that feeling so different in straight people? Or is it that gay people have to keep it secret and so grow divided, with a bachelor's face to the world and a pang like dying inside?

The reason he got such a late start as a lawyer was that Roger lived a whole other life first. During his freshman year at Harvard Law he was simultaneously writing his dissertation for a Ph.D. in Comparative Literature. That work grew out of a decade of Europe and books, the bohemian ramble, complete with beret in one black-and-white of the period. A month before he was diagnosed, we saw a production of Philip Barry's *Holiday*, and Roger laughed on the way out, saying he'd done exactly what Barry's hero longed to do —retire at the age of twenty.

He left Brandeis after his freshman year and went straight to Paris, where he worked as a waiter and flirted with being a poet. The *patronne* of Restaurant Papille was Madeleine Follain, a painter by vocation, the daughter of Maurice Denis and wife of the poet Jean Follain. Roger reveled in all that passionate life of art, and the journal of his nineteenth year, two hundred close-typed pages, burns with the search for the perfect feeling and the words to speak it. When he finally graduated from Brandeis, he returned to Paris for two more years, working at Larousse and Gallimard. Then he took a long sojourn in the Middle East, where his aunt was married to an Israeli diplomat. Once he wandered for weeks through Ethiopia, eating goat around village fires, walking up-country to the monastic caves at Lalibala, till even the guide lagged back for fear of bandits.

Then in 1965 he packed in at Harvard's Widener Library, reading French, reading everything really, till he finally concentrated on the novels of Henri Thomas. He was senior tutor in Dudley House and took his meals at a co-op on Sacramento Street, a chaotic Queen

Anne tenement bursting with Harvard and Radcliffe students, all at full throttle. Roger used to look back on those years at graduate school with a sort of amazement: to think that life could clear you a space to just sit and read! How he savored Harvard, the elm alleys and the musty bookstores, this place that had turned him down at seventeen and left him crying on the stoop of his parents' three-decker in Chicago.

Whereas I had fumbled my way through Andover on a scholarship, too dazed to do much thinking in the thick of an atmosphere that felt as exotic to me as Brideshead. Yet I breezed into Harvard and Yale on half Roger's intellect, with whole hockey squads of my privileged classmates. Four years later I had neither the analytic cast of mind nor the stamina for graduate work. I made a half-wit decision to be a poet—that was the good half—without preparing any sort of career cushion. I ended up teaching at a prep school out of Boston, with the sinking feeling that I had *become* one of the privileges of the upper classes.

But I wrote my poems and papered the East with them. My particular Left Bank was Cambridge in the early seventies, where poets passed through in caravan, some in sedan chairs, some like an underground railway. Parties in Cambridge had totems in every corner—Lowell, Miss Bishop, I. A. Richards ("Is *he* still alive?"), visiting constellations rare as Borges.

In 1974 I was waiting for my first book of poems to come out and generally going about feeling heavily crowned with laurel. Yet the poems seethe with loneliness, the love that dared not speak its name like a stranglehold on the heart. Roger had just completed a year working in public television for a show called *The Advocates*, where bloody Sunday issues were debated hotly by brainy types. The show Roger was proudest of was about gay marriages. He'd been instrumental in pulling the brains together and airing a wildly controversial notion—single all the while, of course.

We weren't kids anymore. We'd been hurting dull as a toothache for years. When we came together as lovers we knew precisely how happy we were. I only realized then that I'd never had someone to play with before. *There* was a lost time that wanted making up in spades. Six weeks before Roger died, he looked over at me astonished one day in the hospital, eyes dim with the gathering blindness. "But we're the same person," he said in a sort of bewildered delight. "When did that happen?"

I related those two lines to my friend Craig (diagnosed 3/2/85) this past Christmas, and he laughed: "But that's what *you* always used to say in Boston. Roger and you were just two names for the same person." Something I don't remember saying, but clearly it was a collaborative theory of ours, rather like the Curies' twin Nobel.

My recollection of the first year of Cesar's illness is in constant swing, a plumb line describing parabolas in sand. He was down to stay over Thanksgiving break, then for a week at Christmas, and it was those visits that brought home at last the physical reality. The biopsy in his groin had left a wound that never healed, not in the whole twenty-six months of his illness. Small black-purple lesions were clustered at the site, and the leg was slightly swollen with edema, though still there was nothing noticeable, not in his general demeanor.

Yet why was he so tired on Christmas morning that he couldn't go with us to Roger's brother's house for breakfast? It is very hard to separate symptoms and degrees of illness anymore. The dozens of cases I've followed since then have blurred the boundaries. Besides, the particular indignities of AIDS are so grotesque, like that endlessly swelling leg, that the general aura of fatigue and accelerated aging are much more difficult to pin down. But the decision to stay home and rest on Christmas morning was a kind of watershed, as if for the first time the illness had moved to hold Cesar back from life.

The emotional roller coaster was in full operation by now, because I know how happy Roger and I were in mid-November, when we stole away for ten days to Paris and Tuscany. I'm sure now that was a conscious decision too, concrete as the roll of sunset pictures from Big Sur. With Cesar sick, a new note of urgency had crept unspoken into our lives. The edge of the minefield is fairly common ground these days among gay men, and many speak openly of doing what they've always wanted *now*—this month, this summer, before they're forty. Neither of us had ever been to Italy in the fall, so what were we waiting for?

By then Roger kept a diary only sporadically, and one night I left mine in a taxi near Saint-Germain-des-Prés. Two months suddenly vanished, and with them whatever I'd dared to speak about the weight of Cesar's sentence. Next morning, for the lark of it, Roger and I went to Gibert Jeune the stationer, near Place Saint-Michel, where we bought blue-cover student notebooks lined like graph paper. At the time I was reading *The Name of the Rose* as a sort of cracked guide to Tuscany, and Eco speaks at the beginning of the *cahiers* of Gibert Jeune. Roger filled only five pages of his, but on October 31 he writes of us sitting by the Medici Fountain in the Luxembourg Gardens in Paris. He's full of "the drag of nostalgia" and remembers reading Gide's *Counterfeiters* on the selfsame spot twenty years before. There's a brief aside to me: "Paul—the book opens with a scene at the Medici Fountain." Then, at the end of the entry: "These spells of fatigue . . . age? some virus? Nothing at all. Time to get up and move on, says Paul."

Where is the pendulum swinging here? Are we up or down? Or do we not even have a clue till a year and a half later, when we will give anything to be back in the swing by the Medici Fountain? Our swift November trip through the Tuscan hills was the opposite of ominous. The disease had brought its scythe down among us now, and Cesar was who it had chosen. So our role, Roger's and mine, was all the more to be brimming with life, enough to spare to keep our friend afloat.

I can see now that my way of interacting with Cesar that whole first year was full-out Positive Thinking, with a slightly frantic nimbus around the edges, like the start of a migraine. If laughter was therapeutic, there were days we could have cured the common cold. He came down to us to be taken care of in a place where he needed to play no games. A week here, three days there, he basked in the house on Kings Road, gathering strength for the stoic solitary fight to keep his life his own. Meanwhile he taught full time, reassuring everyone who knew that he was quite himself, not to worry. Besides, maybe it wasn't even AIDS. This last whistle-in-the-dark notion asked that we factor in his being from an exotic clime below the equator, plus all that sleeping in railway stations with the lepers in India—as if the whole thing might yet resolve itself as a rare but treatable rural fever. The truth is, nobody believed a word of it. But everyone nodded hopefully, as glad as he to go with a second opinion.

At our house he slumped on the sofa in front of the Christmas tree or out by the pool at midday, waiting to joke with Roger and me whenever we put our heads around the door, but otherwise content to lie about. Husbanding his resources. I can see this quality now in myself as I sit on the hill above Roger's grave, or lie in bed writing in the back bedroom, bundled like Colette, resting in between things—and remember, I am fine.

In '83, after two years of rattled disemployment, writing things nobody wanted, I'd sold and written two screenplays back to back and was feeling flush, so I showered Cesar with presents. He was proudest of a cashmere sweater I found in Rome, gray and pink and pale yellow, as if chosen specifically to draw attention off his leg. There's a picture of him wearing it, sitting on the sofa with Rog: my two best friends. They both ragged me unmercifully for being Father Christmas, even as they tore merrily at the mound of packages. Most of the forty people we had over on Christmas Eve knew Cesar had been diagnosed. In these parts he was practically everybody's first case.

I'm not sure how depressed I was. I expect I was compensating madly. Again it is hard to separate out what is general California body lunacy from the frantic attempt to stay healthy. I know I was spending an hour and a half a day at the gym by then, eating lean and fibrous, like a pure soul who will only consume fallen fruit. I took fistfuls of vitamins, cut way back on smoking dope, and otherwise diluted a potentially nuclear case of hypochondria with waves of holistic self-absorption.

How else to explain this peculiar entry in my journal for mid-December, when I agreed to pose in the nude for Jack Shear, a photographer friend of ours. Roger and I collected vintage photographs and enjoyed the work and heated opinions of the younger artists. I'd offered up myself to several of these photographers, whatever project they needed a model for, because I wanted to get inside this medium that riveted me. Narcissus and I are not unacquainted, to be sure, but I also had a confused romantic idea of the moment when Picasso showed Gertrude Stein her portrait and she said, "I don't look like this," only to be informed, "You will."

In a cold studio off La Brea, backed by a huge wave of white paper like in *Blow-Up*:

> *It felt so weirdly natural—the opposite of looking in a mirror, and like being naked at last, perfectly natural, as one wanted to be at 20. It didn't somehow matter whether one looked good or not. One was a man, pure and simple.*

The body as memorial, locked in time—all those *ones*. While Roger sat in a chair behind the lights, leafing magazines and loyally waiting out this peacock show of vanity.

I think that unconsciously a lot of us were beginning to be pierced with dread of deterioration and lonely death, but again it was slow and subtle, the needles more like acupuncture. I felt exhilarated after Christmas week with Cesar, because we had brought the holiday off like a happy ending. Only I didn't countenance any talk of endings. We were living proof that the best of life could go on as before, weren't we? Then on Twelfth Night, as I packed away the decorations, I had a sudden horror about who would be diagnosed *next* year when the box of tinsel came down from the attic. I couldn't get it out of my mind for days. Time itself began to seem a minefield, the year ahead wired with booby traps.

Yet I see in myself—the roller coaster again—a quality that later drove me crazy in other people, especially after Roger entered the lists. I went back to work with a fury. The first six months of '84 find me pulling together a production of a play in New York; doing draft after draft of a screenplay; making a deal to write a thriller for CBS with Alfred Sole. I met in New York with editors to discuss a new novel. My cottage industry was operating full tilt.

Roger's too. After four years with a corporation and a fifth in a small firm, he'd finally decided in the spring of '83 to hang out his own shingle and work as a sole practitioner. He had no idea whether or not he could make a go of it and hustle himself the clients, but he'd had enough of the system. Since he'd paid all the bills during my two bouts of bad harvest, when I couldn't sell a paragraph, I was eager to return the favor. The worst that could happen, I told him, was he'd fall on his face.

But the fact is, he loved the freedom from day one and pulled together an oddball miscellany of clients, from the Downtown Women's Center on skid row to an equity-waiver house called Room For Theatre. A couple of writers, photographers, budding directors . . . but here I am listing all the clients who didn't bring in a penny. His focus was on small businesses and one-man operations like our own. He worked twice as hard as he had in the corporate tower, but his work was all his own and human-scaled.

The brooding, fretful Sundays vanished, when he used to despair with some variant of: But what am I going to do with my *life*? In truth he had come to a point where he could almost laugh at the existential thread that linked him still to the Paris of the perfect feeling. Roger was a happy man with an ache inside about beauty and time, like a character out of Lawrence. There would always be a part of him that longed to be a poet, but having his own practice brought him to a place of delighted engagement and satisfaction that I'd never seen in him before. Besides, he had a poet in his pocket.

There was this unspoken agreement between us: We were both putting in some very hard work for the sake of the longer-term freedom we were demanding more and more of. We'd sit on the front terrace on Saturdays, dawdling over coffee and reeling off places we hadn't been to yet versus those we had to go back to. Perhaps it was implicit now in the gathering dark of the plague that we would try to cram as much in as we could. On the plane coming home from Italy, Roger had turned and said, "Can we go back to Europe in the spring?" I grinned and nodded, and we did it. Cesar was the sword in our lives that proved there wasn't a day to waste.

I wish of course we knew then what little we know now. That the Western Blot had been in place and we could have been tested for antibodies. That the antivirals had been sprung from the pharmaceutical morass. Then we would have slowed down and watched and monitored, the way I have myself. I wish my fellow warriors hadn't lost the first four or five years bogged down by homophobia and denial. When Larry Kramer tells Mathilde Krim in *Interview* about the closeted gay man at the National Institutes of Health who buried the AIDS data for two years, that's when I understand how doomed we were before we ever knew. It will be recorded that the dead in the first decade of the calamity died of our indifference.

Still, it would have taken a lot to slow us down. Our drive to be at the center of the lives we'd fashioned was far too urgent. By now we were starting to know more and more cases. Though none was as close as Cesar, we would hear it murmured about one and another: Michel Foucault the philosopher, this actor, that dancer, all innuendo and secrecy. A distinguished and sweet-tempered producer we knew had been in the hospital for months now, but no, it wasn't AIDS. The disappearing had begun.

There was nothing we could do about those cases, so we anchored ourselves in Cesar's case. I wondered sometimes if he was the first person from Uruguay to have it. The Africa connection was beginning to be in the news, with figures so pandemic the mind rejected the parallel. It was still only five thousand gay men, and nobody knew how long they had, but maybe years. Cases that died in a month were something else; there was nobody fighting.

Meanwhile I was beginning to witness stages of denial I'd already been through, and they left a taste like dirty metal in my mouth. Gay men in the high purlieus of West Hollywood—that nexus of arts and decoration, agentry, publicity, fifteen minutes in a minispot—would imply with a quaff of Perrier that AIDS was for losers. Too much sleaze, too many late nights, very non-Westside. And that's when I started getting angry, because my friend in San Francisco was limping to General Hospital and queuing for hours to see doctors who hadn't a clue. The guesswork of chemo, the leg that kept swelling, the scatterburst of lesions; and still the aerobic crowd was playing *us* and *them*. I saw a split develop in gay men around that time, as people fled into themselves. Gay liberation had only begun in 1969, when a gaggle of Village drag queens drew the line in the dirt outside the Stonewall Inn, resisting police harassment once and for all. Yet the solidarity that followed Stonewall wasn't rock-hard, binding us like the dissidents in Russia. *AIDS* was the jail with bread and water, but there were gay men who would not hear of it. Too much of a downer.

So I would talk about Cesar and explain what he was going through. I wanted to shove people's noses in it. The AIDS jokes began among us, or we adopted them comfortably enough—after all, Eddie Murphy was a funny guy otherwise. The story I want to tell is about heroism and sacrifice and love, but I will not be avoiding the anger. I watched AIDS become gossip, glib and dismissive, smutty, infantile. I gossiped myself. It was sometimes the only way to talk about it, but all the same it's a yellow and disgusting way.

For the time being, however, Cesar's condition remained "stable," a word that would have so many gallows shadings over the next years that finally it would come to mean simply "alive." I remember in the wake of the Chernobyl disaster, or was it the Bhopal disaster, midlevel officials speaking of the situation turning stable. Euphemism, the twentieth century's most important product. For Cesar it meant the lesions weren't dispersing exponentially, the swelling was no worse, and he had enough energy to finish the school year. Staying with us over Easter break, he decided to go ahead with another student tour of Europe, culminating in Spain, a country he considered a serious lacuna in his life's itinerary.

So it seemed the most natural thing in the world that Roger and I should be planning our own next trip, having decided over all that Saturday caffeine that we couldn't go *anywhere* else before Greece. Roger had passed through it once, but knew the place had been beckoning me for years. Besides, our travels seemed to be taking us ever deeper into the Mediterranean, and I promised Rog the next trip would be an overdue return to Israel. Thus basking in the stasis of things, we left for Athens on June 1.

I don't know how not to gush about it. I realize I'm hardly the first to feel it, any more than Byron was, but the moment we set foot in Greece I was home free. Impossible to measure the symbolic weight of the place for a gay man. We grew up with glints and evasions in school about the homoerotic side, but if you're alone and think you're the only one in the world, the merest glimpse is enough. The ancient soil becomes peopled with warrior brothers equal to fate, arm in arm defending the marble-crowned hill of democracy from savage hordes. The source of such heroics is buried very deep—for me it lies in History I at Andover, the stone swell of the athletes' muscles and marathon battle statistics, war after war till it all disappeared.

But you find that your first bewildered erotic connection at fourteen stays with you, since most of the rest of gay history lies in shallow bachelors' graves. I admit the baggage I took to Greece was cumbrous, that I swept across the Aegean at a fever pitch. But I can't begin to say what brought us through the fire without telling this part at a hundred and three degrees. It was the last full blast of sunlight in our life. There is no medium cool for the final pang of joy, no more than there is for the horrors that wait like the Sphinx at the bend in the road.

Others had groaned to us about the smog and chaos of Athens, but we savored every inch of it. The city was brisk with the winds of early summer, and the grass was thick in the *agora*, where Socrates had held forth. We drove to Delphi across the Theban plain, and I wrote in my diary that the day and a half we spent there was the deepest I'd ever plumbed my life. Roger and I walked about with an armload of guides, reconstructing the whole site, from the cleft in the mountain where Apollo slew Python to the Hague-like world court that gathered around the oracle in the temple.

I should say that neither of us could bear the predations of modern religion or the bigotry and smarm of true believers of any stripe. Perhaps we were atheists by default, but the matter of God did not come into the equation of our love. We were, however, the most unreconstructed romantics, fanatic after our fashion as the human waves of Iranian boys who choke the ditches of Basra. And though I have come to be more godless than anyone I know after all this meaningless suffering, I have to admit that day at Delphi—plunging my hands in the Kastalian spring, where supplicants to the oracle purified themselves—I felt more religious than ever before or since.

Yet it was the ancients' religion, with its powers of earth and water, quake and tempest, especially its goddess of beauty and god of the sun. When I ran in the grassy stadium high on the mountain where the Pythian Games were held, their heroes sung by Pindar, I knew I was poised at the exact center of my life. I belonged at last to a brotherhood where body and spirit were one. When a victor at the games returned in triumph to his home, the city wall was breached to show that a place that possessed such a hero required no further defense. In the pitch of the moment it seemed to me that Roger and I and our secret brothers were heir to all of it.

Just before we left, Cesar had said that ancient places "confirm" a person, uniting a man to the past and thus the future. Confirmed was just how I felt by the Greek idea. Hopeless romance, I know: they kept slaves, their women were powerless, they sacrificed in blood. But a gay man seeks his history in mythic fragments, random as blocks of stone in the ruins covered in Greek characters, gradually being erased in the summer rain. We have the poems of Sappho because the one rolled linen copy stoppered a wine jug in a cave, and the blanks are the words the acid of the wine has eaten away. Fragments are all you get. You jigsaw the rest with your heart.

It was the happiest time Roger and I ever spent together. A tourist's route, make no mistake, nothing too out of the way, but the full Aegean odyssey all the same. Three days in Crete, by sea to Santorini, then Mykonos, with a side jaunt to Delos to see the lake where Apollo was born. The latter drained and weeded up now, but guarded still by the stone lions of the Naxians, six of them crouched in the baking sun, staring into the same distance where the statues of Buddha stare.

I look at pictures of Roger taken that week in the islands—the Minoan palace at Knossos, the valley of the windmills high on the mountain where Zeus was born, a wall above the caldera at Santorini. What am I looking for signs of—weight loss, pallor, fatigue? He looks terrific, as dazed as I by the clear light. Sometime that summer I know he began to cough, but ever so slightly, hardly more than a clearing of the throat late in the evening, or was that when things grew quiet enough for me to notice?

I remember a morning in the ruins of ancient Thera, on the lee side of Santorini away from the volcano, the explosion of which in 1500 B.C. gave Plato his myth of Atlantis. The tidal wave hit Crete in an hour and leveled the whole Minoan palace culture. Thera perches on an abutment half a mile above the sea, with a view that goes all the way to Africa, a great sentry post to watch for Persians.

They had an oracle there too, and a gymnasium with an outdoor court at the tip of the bluff where the boys danced naked to Apollo. There are inscriptions along the walls, erotic poems to the boys, though the guidebook wouldn't recite a single line.

This was not even the same galaxy as History I. We were the only tourists there that hour in the dancing court, and we couldn't decide which was our truer ancestry, the boys or the dirty old men. You must discover the pagan on your own; it's not in the books. As to where a man finds his ancestors, the longing is diffuse till you come to a place where they spring full-blown. It had happened to us once before, in a ruined abbey in Provence, north of Arles, standing in a tower and gazing out over the marshes. A dozen white horses were cropping the pale grass. People who travel have dreamlike moments where they borrow time from the past, but it's not out-of-body at all. The echo of the ancient image, warrior or monk, is in you.

Luckily we'd decided to fly home via Paris and London, because after Greece we had to de-escalate and re-enter. A couple of *bonnes soirées* with Madeleine, a day at the British Museum, and on the final leg to L.A. I turned to Roger and said, "We could always go to Egypt next time and *then* go to Israel. Do they let you do that?"

I'm not sure how far I re-entered. Four weeks later the Olympics began in L.A., and you didn't have to be a rocket scientist to keep the connection Greek. Roger went to the diving competition with Richard Ide, our buddy from USC who teaches Shakespeare. Then all three of us attended the final day of track and field, August 11 in the Coliseum. The gridlock crowds that had been expected in L.A. never materialized, and the city was curiously quiet from mid-July on, like Paris in August.

As to whether Greece was a great whirlwind of denial, I note a strangely ashen mood in a paragraph written on a plane July 8, Atlanta–LAX. Alfred and I had been through Savannah, researching a ghost story. On Friday the sixth I'd jotted the inscription off a duelist's grave.

*James Wilde, 22, shot by a man who a short time before would have been friendless without him.*

Was that a gay story? I wondered. Then on the plane home, this disconnected passage where I sound more Tennessee than Georgia:

*It is not a matter of the summer—I mean, not the summer out there, in the muggy breeze off the river in Savannah (not so muggy, not what I thought, not what I wanted), but the summer I used to cry out for and run around in, looking looking. I do believe there is a constant summer in my life . . . till death anyway, till I die or Rog dies or Death starts being everywhere.*

Perhaps the world is always full of portents, as the oracle maintained it was, in every flight of birds that passes. The only thing we could do to hold the fates at bay was to keep our own world full to the brim, or that at least is how I read the magic of that summer's end. As the reassuring postcards arrived from Cesar in Madrid, I cleared space to pick up a novel I'd stopped at a hundred pages eighteen months before. Roger and I decided to produce my play ourselves. Roger was on the board of Room For Theatre now, and he arranged for us to meet with production people and directors on the equity-waiver front. It was a neat close of a circle for us, since *Just the Summers* was about the testing of a marriage, and I'd written it during the most difficult year of our own.

I always hesitate over the marriage word. It's inexact and exactly right at the same time, but of course I don't have a legal leg to stand on. The deed to the house on Kings Road says *a single man* after each of our names. So much for the lies of the law. There used to be gay marriages in the ancient world, even in the early Christian church, before the Paulist hate began to spew the boundaries of love. And yet I never felt quite comfortable calling Rog my lover. To me it smacked too much of the ephemeral, with a beaded sixties topspin. *Friend* always seemed more intimate to me, more flush with feeling.

Ten years after we met, there would still be occasions when we'd find ourselves among strangers of the straight persuasion, and one of us would say, "This is my friend." It never failed to quicken my heart, to say it or overhear it. *Little friend* was the diminutive form we used in private, the phrase that is fired in bronze beneath his name on the hill.

I say all this about names because we were coming fast on Labor Day and our tenth anniversary. We pulled together a party for fifteen, the closest friends we could find who hadn't bolted the city to end the summer on a proper beach. Actor, poet, professor, journalist, shrink, cardiologist—half men, half women, half straight, half gay. In a word, the family.

A couple of weeks before, I'd bought at auction three photographs, and they were by way of my anniversary gift to us both. I admit that Roger's first reaction was very Roger. He shook his head and said we were spending too much money, though he enjoyed the collecting as much as I and loved to walk about the house showing people our pictures. Anyway, these three images were new: a Philippe Halsman casual portrait of Cary Grant and Audrey Hepburn; a Kertesz of an angel on a roof at Chartres with two birds flying over; a Dorothea Lange of a haunted man at a counter in a diner.

I sift these details now because they are so concrete, still here in the house, evidence of all the roads of our lives in the time before the war. Where the love had grown after ten years is not so quantifiable and will never go up at auction, though as I write I'm glued like a tabloid to the sale in Geneva of Wallis Simpson's king's ransom. What is the tenth anniversary in lawful marriages? Crystal? Brass? I somehow never learned such milestones, any more than we ever got eight toaster ovens to mark the white specific day. We still made do with the day we met.

Kathy Hendrix, who was there that night, tells me she recalls with a stab of pain the joy of the anniversary party whenever she thinks of the onslaught of the next two years. Me too. I grope back to it and study the snapshots, ever the scholar of what we lost. Around midnight I toasted Roger with a line from Du Bellay: "*Heureux qui comme Ulysse a fait un beau voyage.*" The words are set in paving stones around the points of a compass, on a crag above the harbor in Nice. Happy the man who like Ulysses has made a beautiful voyage.

To us.

# · II ·

We had lunch at Trumps, end of September, with John Allison, to explore the possibility of his directing *Just the Summers*. John was managing director of the Callboard Theater, a blowsy Forty-second Street space just off Melrose, and he was known for getting the text right, especially comedy. We all got on like instant cousins. The backbone discipline of John's National Theatre training in London shone through all his irreverence. Though he was surely in his mid-forties, he looked about twenty-five. He'd blurted out in a formal oration at his public school in the ninth grade the fact that he was gay, and still bristled with pride to recall the headmaster's rattled dismay. Roger, who'd tracked John down, beamed at the two of us trading shoptalk, and by the end of lunch John was committed. We'd do the play in the winter, spring at the latest.

John would be dead of AIDS eleven months later. He knew his diagnosis when we met him, but was single-mindedly forging ahead and planning new work. He was eager for us to see his production of *Saint Joan* at South Coast Rep, which we thought very sharp and elegant, despite a Joan who seemed to want to lead her army to Glendale. One critic observed that it wasn't up to John's high standards, and a year later I understood exactly the way his mind wasn't quite on it, the vigor not up to the passion.

Over and over I've watched those who are stricken fight their way back to some measure of health and go on working—those who are not let go, that is. Perhaps the work is especially important because AIDS is striking so many of us just as we're hitting our stride at work. I mean of course the American AIDS of the first half-decade, before it began to burgeon in the black and brown communities. Most of the fallen in our years were urban gay men, and most of these were hard at their work when the symptoms started multiplying and nothing would go away. They wanted to hold on to their work as long as they could.

Roger was no different, and neither am I. With him gone, there is just what work I can finish before it overtakes me. Again there are friends all around me, meaning well, who say I don't have to feel so cornered: New treatments are coming down the pipeline every day; the antivirals data looks better and better. Et cetera. Perhaps it is just very human to want to die with your boots on. I don't know if that's a cowboy or a combat metaphor, but both are perfectly apt.

After six years in the house on Kings Road, we'd fallen into a pattern of optimum tranquillity. Our most consistent time together—though we'd pick up the phone all day and call—was evenings seven to eleven. Some tag end of the workday might spill over here and there, but we usually ate at home on weeknights, did some reading and went for a walk. My whole life is like one of those weeknights now, plain and quiet and, here in the house at least, close to Rog. A stucco thirties cottage high in a box canyon above the Sunset Strip. There's a view of the city lights through the coral tree out front and between the olive and the eucalyptus across the way. In square footage it's about the size of a two-bedroom on Seventy-ninth and West End, the sort people kill for. Out back is a garden court shaded by Chinese elms and a blue-bottom pool that catches the sun from eleven to three.

There was always a sort of double clock to the evening, because Roger was asleep by midnight, never a night's insomnia, and I didn't go to bed till three. I typed like a dervish once the phone couldn't possibly ring. But I'd usually loll in bed with Rog for a half hour—Ted Koppel too, if the issue was ripe—and he'd nod off curled beside me, the two of us nestled like a pair of spoons. By way of trade-off I'd be half aware of him getting up at seven-thirty, padding about while I burrowed in for the morning, to rise at eleven like Harlow. Between us we covered the night and the morning watch.

I realize now how peaceful it was to be writing while Rog lay asleep in the next room. I can't describe how safe it made me feel, how free to work. I think mothers must feel safe like that, when it's so late at night you can hear a baby breathe. We had gone along this way for so many years that when I had to do it for real—watch over him half the night, wake him and give him pills, run the IV, change his sweat-soaked pajamas three different times—it never stopped feeling safe, not when I had him at home. In the deep ultramarine of the night, nothing could really go wrong, and nothing ever did.

In October we managed to get away to Big Sur as usual, though Roger was working time and a half, having taken on several projects for other lawyers. I was back into my novel about a nymph and a loveless man, very Aegean. At the same time I was steering through the Hollywood jabberwocky a project called *The Manicurist*, a comedy for Whoopi Goldberg. The trip up Highway 1 on an aching clear

Sunday morning was our first long ride in the black Jaguar, a vehicle we had acquired by default, after my Mercedes was stolen at gunpoint on the Strip.

Just before we left, Cesar was down for half a weekend. We tooled around in the Jag, and he seemed in fine shape. Relations had suddenly gotten tight with his friend Jerry, an antique dealer he'd known for years. They might've become an item once, but the stars were crossed. Now they were spending a great deal of time together, no term of attachment required. "Buddies" is what has evolved in AIDS parlance for the bond between the mainstay friend and the one in the ring doing battle. Jerry had clearly come forward to take that role, but Cesar wasn't acting as if it had anything to do with his illness, properly so. The bond between them had its own sweet Platonic tang, and Cesar was thrilled to have somebody else to talk about.

It was the last time he would play down the swelling leg and the lesions, regaling us with tales of Spain. All summer long, when anyone would whine at me about some benign indignity of daily life, I'd stare and say, "Cesar's in Spain," as if to sting them with the challenge. What the fuck was their excuse? Now, in October, Cesar remarked to us offhandedly, "You know, I've traveled enough now," but it didn't seem morbid or ominous at all. He specifically meant he'd rather come stay with us when he had a little vacation. Enough of the world out there.

Besides, we were clearly holding the line. A year into his diagnosis, and he'd still never been in the hospital. His doctors kept telling him there were drugs in the works nearly ready for testing. Research was galloping. Keep taking the chemo, they said, even if it didn't seem to be doing a thing. I don't recall seeing his leg naked during that brief visit. We'd dropped the fiction of the rural virus in the terminus at Benares, but only because AIDS was proving manageable. Management skills were what we needed now.

The obverse of this optimism was the hair ball of fear at the pit of my stomach. I'd convinced myself by this point that I was more than likely in the direct line of fire. I can't say what was hypochondriacal here. It was certainty born of dread: The glands in my neck and armpits were no bigger than almonds; they didn't hurt; they were nice and soft. Moreover, they didn't appear to be growing, but oh they were most definitely there.

My doctor's little speech about them, reiterated for two or three years now, came down to the same bland assertion that they could be anything. Dozens of things make the lymph nodes swell—stress, for instance, the blanket diagnosis of the age—but now the news was getting very specific about the lymph nodes being a flashing amber sign. *Pre-AIDS*. We still had that word then. Certain gay men I knew, in fact, were becoming obsessed with the notion. How deep exactly did *pre* go? Could you see it in a person's face? And how much time before *pre* burned down like a fuse on a keg of powder?

I tried not to talk nonstop about it; it sounded vain even then. I simply redoubled my efforts to mount a holding action. Lifecycle at the gym, vitamins, writing in bed, monitoring my almonds like a sort of DEW line. I vividly saw the process as a struggle to keep it from breaking through—a wall of water behind a dike, or the mangled son pounding on the door in Kipling's "The Monkey's Paw." "Breakthrough" was not then commonly used to describe the onset of full-blown infection, but the word has just the right edge, chilling and paranormal, like the breakthrough of alien life out of John Hurt's belly.

I knew all the warning signals now, rote as the seven danger signs of cancer that I carried on a card in my wallet in high school. Did I think I'd forget them? Night sweats, fevers, weight loss, diarrhea, tongue sores, bruises that didn't heal. None of the above. But I'd run through them every day, examining my body inch by inch as cowering people must have done in medieval plague cities, when X's were chalked on afflicted houses. I didn't even want to eat Asian food anymore, because it shot my bowels for a day after.

Any change, any slight modification . . . even a bruise you remembered the impact of, you'd watch like an x-ray till it started turning yellow around the purple. KS lesions do not go yellow. They also do not go white if you press them hard with your thumb. A whole gibberish of phrases and clues was beginning to gain currency. A canker sore in the mouth would ruin a day, for fear it was thrush—patches of white on the gums or the tongue. I read my tongue like a palmist before I went to bed at night.

In none of this paranoid fantasy did I have the slightest worry that Roger was at risk. I hadn't forgotten the flu in '81, or the assault of the wrongheaded drugs for amoebas. There were shakes and fevers that winter, and for a week or two Roger would break out at night in hives the size of silver dollars. It had been an awful siege, but that was all three years ago now. Never a complainer about his health, he didn't mention losing weight till the end of November, and even then it was only a couple of pounds. He was tired at night, but a wholesome kind of tired, with a long untroubled slumber from twelve to seven-thirty like clockwork—what the French call *le sommeil du juste*. And his cough was still such a minor two-note matter. He'd be putting on his pajamas, and I'd turn from *Nightline* and tease him: "What are you coughing for? Stop it." That was how ordinary it sounded.

Is that denial? If it was, it was warring in me with a doomed acceptance, as I struggled to figure how I would bear the sentence myself. Late at night I'd walk in the canyon and think about Roger watching me suffer. I was already riddled with guilt: None of this would be happening if I'd never had sex with strangers. I suppose I felt there was something innately shameful about dying of a venereal disease. All the self-hating years in the closet were not so far behind me. And any brand of shame lays one open to the smug triage of the moralists, whose vision of AIDS as a final closet is clean and efficient as Buchenwald. Of *course* we didn't deserve this thing, but how do you go up against them when you're suddenly feeling wasted by every lost half hour in bed? After all, the very qualities that used to recommend such aimless sport were its junk-food suddenness and its meaning nothing.

My therapist, Sam Dubreville, reeled in every tortuous loop of self-flagellation. All right, so I couldn't deny the dread. The menace was real as the man with the .38, swinging it wildly back and forth between my head and the gray Mercedes. It might be true that all of us were trapped by the careless time before we heard the first siren. But the disease wasn't drawn to obsessive sex or meaningless sex. Sex itself, pure and simple, was the medium, and the world out there was ravenous for it. Straight and gay alike, they wailed like Patsy Cline, rubbing up against their home screens. Don't personalize the illness, Sam said, don't embrace it with obsession. Live now, in other words, sobered and alert. Relish the time Roger and I are whole, because something is going to beat the door down someday. *Live now* sounds simple enough to be carved on the temple at Delphi, except they preferred to chisel instead: *Know yourself.*

I did what I could with my panic, riding the energy like body surfing, turning its intensity to consciousness of now, where Roger was. I started making plans for Thanksgiving and Christmas before Halloween. In this regard Ned Rorem recounts a crystalline remark of Jean Cocteau's. When asked what one thing he would carry away from a burning house, Cocteau replied, "I would take the fire." There's something in there about the fire of inspiration, but I choose to see it the other way, carrying out the fire to spare the house. I made time happy. I worked at it. And when Roger and I would saunter through the County Museum or run out late for Häagen-Dazs and a stroll in Boys' Town, the thought of him all alone without me—alone in our house, in this city where he came to be with me—would vaporize like a bad dream.

On November 7 we had dinner with Rand Schrader, municipal judge and past president of the Gay and Lesbian Community Services Center, which mouthful was the premier channel of local good works for those on the bus. The Center was a national model for a kind of gay sanctuary for the troubled, lost and burned. It was the yearning to be free to love that appeared to bond us as a people, a bond that turns out to be as strong as land or language. Roger and I were part of a growing core of gay professionals who helped to fund social services for runaways, alcoholics, the banished of all persuasions. The Center occupied a building that had the motel air of cheap dentistry, but proudly wore its long name ribboned across a lintel on Highland Avenue.

We were together to talk about the Center's annual dinner, where Abigail Van Buren was being given the Ambassador's Award. I was to write introductory remarks for Julie Harris, who would make the presentation. What you have here is practically the textbook definition of thankless writing. But I had a particular memory of reading *Dear Abby* as a kid. In a straight New England town, out of reach of therapy, Abby's column in the Hearst rag was the only consistent forum for the radical notion that gay people could be happy despite the hate and discrimination. *Love and let love* was more or less the way she put it. The effect is incalculable, to finally hear somebody say it isn't wrong.

Rand was probably Roger's closest friend in L.A., a passionate, unguarded man who was out in every quarter of his life, with a decade of therapy behind him to prove it. Roger and I always sponsored a table at the Center dinner, though Roger was the one who gently badgered our friends to fork over the two-fifty for the ticket. Gay people have had to be taught to take care of their own, having grown so accustomed to taking care of themselves. I couldn't bear the drone of self-congratulation at these affairs, but Roger said it was boredom in a good cause, so we always went and I helped with the speeches—shortening, shortening.

It is in fact very moving to hear Julie Harris read your lines, and she pulled off every laugh and stirred the place too. AIDS was a growing subtext at any community gathering now, but it was salutary to recall that as a people we were still making progress, countering hate. My first year as a teacher, twenty-two and stuck in a prep school run by Dickensian colonels in Connecticut, I had a student called Styler, by turns diffident and shyly charming, working to please, wouldn't swat a fly. I was in the closet and never thought twice about him, until three years later when he killed himself, and his sister wrote to tell me. Oh, I thought with a knot of hopeless sorrow, so he was gay. I hadn't thought to help him, because I couldn't even help myself then. Anyway, I knew we were sitting in tuxes at these gelid chicken dinners for Bob Styler's sake.

The night before Thanksgiving, Cesar arrived, and within five seconds of lurching in he fell in a chair, near hysterical. Not crying or raging, just letting it all tumble out—the circular talk of the doctors, the brave front at school, the pain and rot of the swollen leg. The dog went up to sniff at him in the chair, just the regular sort of canine radar work, and Cesar cried: "See him smell it? I can't stand it anymore!"

*It* was his leg.

We just let him talk till midnight, apologizing even as he spilled it—how could he bring this misery on our house? We reassured him over and over, gently easing him off the panic, sorting out the players around his Lear, finally comprehending how much he'd been bottling up ever since school began in September. Still fighting not to depend on anyone, he who always seemed to have half a dozen shoulders to prop up all his friends.

Through the whole four-day weekend we forced him to rest—allowed him to, really. Ten, eleven hours a night he slept, and seemed to get stronger and laughed again. But the leg was really repulsive now, twice its proper size, raw and mottled, the lesions clustered at the groin angry and blood-blistered around the wound, which sagged open and wept. It had been that way for a year now.

Yet at Thanksgiving dinner in Robbert Flick and Susan Rankaitis's studio, his social grace was mesmerizing as ever. I could see how drained he looked, his hair beginning to thin from the chemo, circles under his eyes. He told Roger and me how he would sneak naps in the storage closet off his classroom, curling up on the floor during his free periods. But over dinner he scintillated and coaxed out people's travels, tossing a few doubloons from his own. And when it came to books and films, the irrepressible passion was undiminished, pleading with people to read *Sentimental Education* or go rent *Gilda*. The dinner table was Cesar's stadium, no doubt about it.

You walk in a daze through days like these, working to keep life normal for your friend, trying to give him a respite; that much you can still do. But the frailty of life has got its hooks in you, and a lot of the cheer is hollow and ventriloquized. My journal says we laughed, though. The three of us went to *The Terminator* on Friday night and loved it. The first good movie since *The Last Metro*, we decided. Or maybe what we said was that Arnold Schwarzenegger was the Catherine Deneuve of violence. That was the way we jockeyed culture, keeping it aloft like balls in the air.

Saturday night we had about ten people in for the evening, including Cesar's friend Jerry, who'd driven him door to door and was staying the holiday with his family in L.A. Cesar was visibly pleased to have his companion near, though Jerry himself was a wreck. His mother was dying of cancer, and he'd been out of work for months. Yet he seemed so loving of Cesar, so glad to be with him and hear him talk. There was no doubt that the two of them were racking up lost time like pinball, no matter how little there was to play with.

And with the party in full swing—just the size Roger liked, conversation one on one—we seemed for the moment safe again. We would fashion our own reprieves. In the good hours, I still had an almanac faith that the proper doses of rest and love would bring things back to stasis. Before we all went to bed we struck a deal whereby Cesar would come for a week at Christmas. Meanwhile Jerry was picking him up early Sunday morning for the long drive home, and Cesar promised he'd peek in to say good-bye before they took off. On Sunday I woke to him lightly touching my shoulder and saw he was kneeling beside our bed. "Good-bye, darling," he said, smiling through a glaze of pain.

"What's wrong?"

"I just threw my back out." He was literally bent double to the floor, trying not to gasp.

Cesar's back. I remember when it was the worst problem we had—Cesar's bad back, mine, Roger's, in descending order of magnitude. (Craig, last summer: "Remember when all we worried about was whether the melon was ripe?") By dint of the vaunted Williams exercises, Cesar had brought his back into supple shape again, but the weight of the swollen leg threw him off balance. When Jerry arrived, Cesar could just stand up and hobble down to the car, me with his kit bag. I stood in the ludicrous morning sun, trying to wish him Godspeed, and he just kept shaking his head.

"They've got to do something," he said. "I can't go on like this."

He meant the doctors at San Francisco General, still convinced they could alchemize the optimum dose of chemo. I didn't know what the drugs were; it was the last case before my pharmaceutical residency at UCLA. In addition, we were getting all our information about the treatment from the patient. Despite Cesar's fluency in four tongues, there appeared to be a language barrier here, because he couldn't seem to tell us what the course of the medication was or what the data said. We were intellectuals. We needed an idea.

I still didn't fully understand—nobody I knew did—the difference between KS and infections like pneumocystis. Three months later we would learn that the scythe fell one way or the other, lesions or the lungs. This was the crude half-picture at the end of '84, before we knew about lymphomas and the brain. Men with KS were seemingly the lucky ones, because they appeared to have the strain with a slow progression. But I could see myself now that KS was something a good deal more dire than skin cancer, its lesions rooted deeper than bruises. They looked like exploded blood vessels under the skin, and sometimes they boiled to the surface like stunted orchids.

We were helpless, Roger and I, six hundred miles away. We felt like calling the doctors ourselves, to affirm somehow that we were Cesar's family here in the States. We worried that he was going to a hospital already bursting at the seams with AIDS cases. More to the point, Roger had always been alert to Cesar's difficulty with father figures. He was a charmer, not a demander. We could imagine him being passive with his doctors, politely letting them experiment. You don't need cautionary tales like the *Titanic* to know how many survive in steerage and how many on the boat deck. Unless you have a private doctor with privileges, which is another way of saying you'd better have money, you are lost like Hansel and Gretel in the system's beige-flecked corridors. The peaks of insurance pale beside this Everest of a condition.

It is also imperative to have doctor friends to run things by, to provide little crash courses in hematology. Even a nurse can steer you around so you know which way is north. Fortunately Cesar had Lucy, a nurse who worked part time at a high-toned white-bread hospital, the Pacific Heights wing of medicine. It had taken fifteen months, but finally Cesar said *Help*. And as soon as he went into Lucy's hospital, he suddenly wasn't another dumbstruck face in the Kafka crowd at the clinics. They put him on intravenous chemo right away. The bullet wound from the biopsy was part of the KS, light-years more evolved than a mere bruise. Cesar's new doctor told him the clinic had been treating him with a dose of chemo that was far too low. Yet the doctor wasn't blaming anyone, since he knew the free clinics were swamped like a field hospital at Verdun. The good news was that the higher IV dose would push the cancer back, and Cesar would be out and down for Christmas as planned.

There's a great harp-string moment in *A Christmas Carol* when Scrooge leans out his bedroom window after the nightmare—George C. Scott did anyway, but maybe this was a rewrite—and begs to know what day it is. "Why, Christmas," a boy tells him, stunned that somebody wouldn't know. And Scrooge clutches his hands and gasps: "There's still time!"

That is how I went around for the next four weeks, building Christmas like a fortified town, even though I was down with a cold within hours of Cesar's wrenched departure after Thanksgiving. A cold that lingered: I slept a week in the guest room so I wouldn't pass it on to Rog. Even that unpassed cold gave me a sort of subliminal relief, for it seemed to prove that Roger's resistance was strong. He was swamped at work as a blizzard of legal matters converged on the end of the year. But we still had time to spare, and I meant to spend it on our friend. Roger, who'd grown accustomed to the jingle of bells in December, may not have noticed the pitch of frenzy that attended these preparations. I stopped short of stringing popcorn and cranberries, but just barely.

Meanwhile I got lucky the first week of December. After six months of serious hustle—being "encouraged to death," as Pauline Kael puts it—I managed to put *The Manicurist* into development at a studio. "Development" is the Hollywood term for suspended animation, but at the beginning at least there is the deal. I knew that once the negotiating was done I'd have my next year's work cut out. I even managed not to worry how I was going to write jokes in the fifth year of the plague.

I spent my afternoons shopping, a blur of packages mailed to the four corners, mounds of presents for Cesar and Rog. Cesar had been with us for the holidays half a dozen times, falling into a cooking mania that went on for days before Christmas Eve. I knew how Cesar would glamorize his visits to L.A., especially back up north, where he found the curled lip of superiority as to the lowness of L.A. deliciously provincial. Nobody understood the provinces like Cesar. He had a raft of Madame Bovaries in his life, one in every port. So I determined to make him a perfect week, a fresh doubloon to fling on the table when people acted as if there was nothing real about him anymore except his diagnosis.

I know I saw John Allison during December, because I remember telling him how undone Roger and I were by Cesar's November visit.

John nodded gravely, explaining he had a close friend who was ill. Then, as we talked about scheduling for my play, he said something very odd: "Of course, I could always decide to toss it all and take off."

"Take off where?"

"I don't know. Doesn't really matter. Maybe I've had enough of theater and all this la-de-da."

Did any of that shiver through me like an echo? "I've had enough of traveling," Cesar had said. Frankly, I was more concerned just then about my play—that I might have a flake on my hands who'd disappear in the middle like a pouting Prospero. John reassured me with an easy laugh that we'd mount my play first, no matter if he ended up in Fiji. But he might have a musical coming along, so we probably wouldn't do *Summers* till late spring. That was okay by me. I figured I'd be up to my neck in *The Manicurist* for the next several months.

Chris Adler, the New York composer, died ten days before Christmas. I note in my journal that Chris and Cesar had both had to suffer "the grossest misdiagnosis." Chris had been dying by inches for six months, bombarded with drugs for lymphoma, at one point requiring removal of his spleen. Yet at no time would anyone call it AIDS. I was close to a heartsick friend of his, and it didn't seem to be a case of stonewalling the truth. Rather, we were still stuck with the CDC's narrow definition. By the time Chris died—at twenty-eight, his family of strangers circled about him, his lover banished from the room—there was no doubt in my mind. It was the first time I wondered how many died and had never made the CDC list, which was hovering now around seven thousand.

I also recall thinking, when Chris's spleen was taken out, about a psychologist I knew whose lover had died in 1980 of liver cancer. That one had struck me as curiously coincidental with somebody else—a best-selling novelist, in the closet of course, who'd died of a fast liver tumor around the same time. And now that I thought of it, a mad and gaudy screenwriter who rode high for seven and a half minutes, with a sweet tooth for porn stars, had also withered and died in the fall of '80, again of liver cancer.

Were they all drinking the wrong kind of vodka? Or was there something we weren't being told about the organs? There was growing frustration—rage in New York—as to what we were and were not being told. Was anybody pooling this data? Sometimes you felt that your own journey and your own circle would give them the full etiology of it, if they would only factor in all these horrible coincidences.

One night in mid-December, Roger came home in great distress from yet another dinner, Lawyers for Human Rights. Rand Schrader had told him that one of the group's officers was dying of AIDS and his family didn't know. They also didn't know he was gay. This was our first encounter with the double closet of the war, the *Early Frost* division.

Roger processed the constant upheaval differently than I did. He anguished for people in pain, moved as he was by solitary lives and the cruelties of fate. A month after I had met him in '74, I came by his Sacramento Street apartment, and he was listening to *Kindertotenlieder*—Mahler's songs for dead children. We were in love, and I was more of a mind to listen to Linda Ronstadt. But if Roger internalized the tragic, it wasn't by way of suppressing it; he could weep openly too. He simply contemplated more than I did what people went through, while I got manic and whipped myself up to *do* something. That's why it's harder to piece the whole of Roger's inner story in those months, for I was furiously acting out, over the top and full speed ahead. Being as we were the same person, happily it all balanced out.

Cesar arrived December 23, spirits high. His back was better; the chemo had checked the cancer for the present; and he felt enthusiastic about his new set of doctors. He was ready as ever to cook up a storm. To our friends who came over on Christmas Eve—last and best of our parties—there was something enormously comforting about seeing Cesar going strong a year later. He was utterly himself that night, the vivid exhilaration fine as silk, for he had more to celebrate than we did. He had pushed the enemy back. The border was barely secured, the truce uneasy, but here was a man returned from the front lines.

During the next few days I orchestrated the time so he could appear and disappear to rest whenever he needed. I knew the rhythm now of what he could do and what it cost. During the down times I'd sit on his bed, which looked out on the pool, and talk the past with him back and forth, never tired of analyzing the strange behavior of the shadow folk who still lived in the closet. When his strength returned we'd go find Roger and off to the next party, where we'd scarf a plateful, make the rounds, then duck out and home to open another present.

Roger, Cesar and I: the Chicago Jew, the Uruguayan lapsed RC and the hollowed-out Episcopalian lost at Delphi. I used to tell the two of them Scrooge's nephew had the proper text for Christmas:

> *The only day I know of . . . when men and women seem by one consent to open their shut-up hearts freely . . . as if they really were fellow-passengers to the grave, and not another race of creatures bound on other journeys.*

I bore no particular animus toward Baby Jesus; he was welcome to the Sunday side of it all. Our own celebration had its fix on the year's end, druid-like, tuned to the solstice glow of pinlights on a Douglas fir. That is to say how it felt in the time before the war, years that ended gliding into a sunset harbor, the three of us out on deck watching the sun sink in the wrinkled Pacific.

In '84 the celebrating had to be danced on one leg. I can see Cesar in his nightshirt, stretched out on the sofa with his whale of a leg, so visibly glad to be doing nothing as long as he was here with us.

I remember one morning he called his mother in Montevideo. He bragged about the party and catalogued all his presents, as if to make her see that things were the same as ever. She told him she felt relieved whenever she knew he was in L.A. I can't imagine what her picture of Roger and me must've been, but I'd trust her version of us sight unseen. I never even wrote to her after Cesar died, because Roger and I were on the beach in Normandy by that time. I can't think who might have her address. How can a man be so dispersed? I guess what I'd like her to understand is that the past he always bragged about was true. The melon was ripe for years on end.

And I've only begun to understand that serene contentment of his, as he lay prone on the sofa restoring himself. It makes me wonder: Was all that hyperactivity of mine—choreographing our coming and going, engineering flashes of how it used to be—was it all beside the point? Somehow it's hard to accept that a man can be totally happy just lying about, no regrets despite the ravages. I was so grief-stricken for Cesar I couldn't slow down long enough to see. Yet I know that happiness all too well now. I saw it exactly the same in Roger a year later—the respite between sieges, delirious to be home, and the preciousness of lazy hours when weeks are life and death.

Diogenes sat in a square in Athens with nothing but the clothes on his back and a tin cup for dipping water from the well. Then he saw a beggar poorer than he drink water out of his hands, and Diogenes tossed away the cup. This burning away of the superfluous, the sheer pleasure of an ordinary afternoon—does anybody ever get taught these things by anything other than tragedy?

I don't remember what sweater I gave Cesar that year, but I made him a present of an hour with Sam. "Just a quick fix," I told him. Sam of course had followed the case from the start. He ran through some relaxation and visualization techniques, did the Cook's tour of Cesar's tangled feelings, and laid out some options about support groups available in the Bay Area. For it seemed to all of us now that Cesar was going it too much alone. I know this sounds crude and intrusive relative to the stately pace of therapy, but so much gay psychology these days is crisis intervention and burnout work. There are very few fine points left when people are screaming in clinics and shutting out friends and leaping thirty stories roped at the waist with their lovers.

Yet if I was a great support to Cesar, there was still something magic I sought in return. "What if I get it?" I'd ask him; that is, how would I bear it? Would I be as tough and noble as he? And he always replied with emphatic resolve: "You're *not* going to get it." I'd practically be coaching him for the right answer. And perhaps he had a corresponding need as well, to believe the nightmare would stop with him, that he somehow bore it for all of us. A martyrdom of sorts? Well, whatever keeps you going. Sometimes he would hoot with disdain as he spoke of one self-obsessed friend or another, locked to the mirror like a dry Narcissus. He knew who could take it and who couldn't.

It wasn't all verbalized. Perhaps I've thought too long about how we spoke when we knew so little. But the paradox was this: I had the courage to face the possibility of my own illness because my loved ones kept up the litany that it wasn't going to happen. I took a similar comfort in the fact that Roger wasn't obsessive about AIDS and didn't go ice cold whenever he saw a funny bruise. Yet if Cesar projected a glimmer of magic for my sake if not his own, I also recall him saying late one night, with an ashen finality: "When are they going to realize they have to stop?"

Stop casual sex, he meant, and he meant gay men. When he said it I found the remark far too extreme, even as I gazed at his purple leg, which now gave off a sweet and sickly smell like burnt flowers at the site of the open wound. That is the smell of dying to me. And the spectral fiat—*they have to stop*—has stayed with me now three years, till I see that he meant everybody. And for all my loathing of the holy lies of straight religion as to love, I agree with him now. If everyone doesn't stop and face the calamity, hand in hand with the sick till it can't break through anymore, then it will claim the millennium for its own.

He left on the morning of the twenty-ninth, patting me half-asleep before he took off with Jerry. The Platonic interlude between them was dwindling down. Cesar didn't have any room for Jerry's despair over his mother. A few weeks later, when Jerry heard his own T-cell numbers were in the normal range, indicating healthy immune function, he exulted to Cesar about it. Cesar took offense and began pulling back. The clearest thing I remember him saying over Christmas came at the end of a long self-critical talk about his failures of the heart, the choosing of people who wouldn't love back. "At least I picked the right friend," he said, waving an arm that took me in and the room by the pool and all our laughter.

When I got up that morning, Roger was out on the front terrace. He'd had an early breakfast with Cesar and then watched him down to Jerry's car. "Oh, that poor man," he said now, choking with pain and a wave of tears. I stood there staring dully out across the milky city awash with the old year's sun, and I didn't know what to say to soothe him. The furious dance of the perfect week was over, and now the terror of what lay ahead came back in its full blankness.

I went in and spent a demented half hour airing and cleaning the guest room, stripping the bed to the bare mattress and wiping everything down with ammonia—a perfect frenzy of prophylaxis, almost a phobia. Guilty and vaguely appalled the whole time, as if I was secretly abandoning my friend. Scared shitless too, because what if everyone was wrong about the virus after all? Maybe it could cling to a pillow slip where a fevered head tossed restless. And if the sweat soaked through to the pillow, did you have to throw that out as well?

We are not just talking about the sterilizing of glasses here. Every second day while he was with us, there was a shopping bag full of soiled paper towels, rank with suppuration from being padded against the gash in his groin. The trash bags massed at the curb that year were a weird amalgam of bandages and gift wrap. Thankfully, that was the last time I was ever possessed by this particular madness, but it's why I have such an instant radar for the bone-zero terror of others. Those who a year later would not enter our house, would not take food or use the bathroom. Would not hold me.

The only thing I could think of to lift the gloom that morning was to get us out of the house. So we went to Michel Richard, where we always ended up on winter Saturdays for café au lait and croissants. Six little spindly tables tight to the sidewalk on Robertson Boulevard: Paris if you squinted hard enough. We needed the ballast of normalcy. And it nearly worked, because here we were, limp with an unspoken gratitude that we at least were spared. Then a voice with a nasal whine said, "Hi," and I turned, and there was Joel.

Joel is a couple of different stories, only one of which is AIDS. He was thirty-four then, a pumped-up former actor given to facials and Melrose threads. When I met him, in the summer of '81, he was writing a play, and I was two feet from the brick wall my career was about to smash head-on like a runaway train. I had a tortured six-week fling with Joel, being pushed away with one hand while the other ran dialogue by me. Like all obsessions, of course, it was lunacy. But even when the affair had died of my own misery, the seduction of being a mentor held. I filtered every line of Joel's play through my own assaulted heart, and the only good that came of the whole mess was the spur it gave me to write one of my own. *Just the Summers* was about Joel and Roger and me, and I couldn't begin to be finished with him till I finished it. Though Roger and I had weathered and survived it, the dead end I was in for months was very hard to come home to.

In the years since, Roger had come to see Joel as childlike and floundering but not unamusing. From that awful summer on, Joel had been hooked up with Leo, a kind and simple man who worked as a caseworker for the Feds. By way of putting it all to rest, Roger and I maintained a certain guarded acquaintance with them, but I hadn't actually seen Joel in nearly a year. What shards of the wreckage remained were like shrapnel the doctors do not feel the need to remove; very deep in the tissue, aching slightly on a rainy day.

Yet I stiffened the instant I saw him that Saturday noon, because the wound of Cesar's pain was still so fresh. There was a group of people waiting for Joel farther down the sidewalk, so the encounter couldn't have lasted thirty seconds. But as he sauntered off, I thought—I wrote it this way in my journal—"Why is my friend sick and this asshole is so strapping well?" Needless to say, I'm not proud of that. The free-floating rage with a hex in its tail, almost wishing the horror on others, is as annihilating as any feeling I've ever had. I knew it was wrongheaded even then.

Besides, Joel and Leo had been through six months' terror of their own, as Leo had been plagued by a set of symptoms he couldn't shake—sinus infection, rashes, fevers. The doctors assured them all along that it simply couldn't be AIDS; Leo just didn't fit the profile. What I didn't know on that Saturday was that Leo's low-grade fevers were getting worse. The day before we saw Joel outside Michel Richard, he'd filled the tub with water and fumbled a razor across his wrists. He didn't get much further than breaking the skin, he admitted as much, but the feel of overload and the rope's end were real enough. Joel had been hospitalized overnight on suicide watch, and the group he was walking with on Robertson was an outing of the self-destructed, trying to make do with one day at a time.

Nothing was simple anymore. Even if it used to be complicated, the complication wasn't AIDS, and now it always seemed to be. Still it was all unfocused, isolated cases, maybe eight hundred in L.A. now. There was minuscule coverage in the local press, and nothing approaching a citywide welling of fear and protest. The most constructive thing anyone seemed to be doing was avoiding all travel to New York and San Francisco, which were now perceived as "over," not to mention a bummer.

Thus I was all alone in my rage over Cesar. I didn't even know how to speak it to Rog, though I'm sure I'd already begun screaming at bureaucrats on the phone and erupting in major outbursts while standing in line at the post office. These three years have taught me that fear—terror, that is, with a taste like you're sucking on a penny—is equal parts rage and despair. The panic makes your brain race so fast that the yelling spews like poison food and the blackness flattens you, without any back-and-forth like day and night, not even any contrast. You are up and down at the same time.

So if I've neglected to mention that Roger was still pretty beat at night, complaining again that he'd lost a few pounds—only three, how could that be anything?—it's because his minor frets were in a whole different league from the general nightmare. I did as much complaining about the malingering of my cold. Roger's late-night two-note cough had gone on so long that it actually made me feel secure. Whereas Cesar wasn't waking from the bad dream.

The year did not end neatly, that much is clear. But years are so ingrained in us—the number of our own, the numerology of history—that we cleave our lives and make our resolutions on the cusp between last day and first day. Which reminds me of another toast: In '75 on Sanibel Island at New Year's, we split a split of champagne at midnight on the beach. We ended up pouring half of it in the sand, preferring the star-shot sky to wine, in the process spilling libation to the gods. And to show how little foreshadowing there was, even on the jamb of '85, I stare as if from a receding train at the final entry for '84. Despite the shell shock after Cesar and the thundercloud statistics, I'm "ready to start the year. R was reading his old journal from 1980 tonight, and we roamed back over things."

I can see us so vividly side by side in bed—reading, dozing, roaming—always coming round again to that evening anchorage, no matter if the day had been a hurricane. It would all begin to accelerate very fast now. Compared to the bend in the road ahead, this last stretch—Thanksgiving to Christmas, all for our brother below the equator—was sweet as a harvest picnic. At the time I thought there were no more layers of innocence to peel. Things couldn't be worse, I'd think sometimes, and that was to *calm* me. "No worst, there is none," goes the line from Gerard Manley Hopkins that would toll out of nowhere in my head.

I cannot say what pagan god it was, but I'd gotten in the habit, last thing at night, of praying: *Thank you for this*. I'd be tucked up against my little friend, perfectly still, and thanking the darkness for the time we'd had—the ten years, the house, the dog, the work. I did, I counted my blessings. Praying for more, of course. Willing—at that point anyway—to make my peace with the infinite, that our life should stay exactly as it was, nothing but nights like these. I knew what I had and what I stood to lose. I held it cradled in my arms, eyes open even as I slept. The night watch from the cliffs at Thera, clear along the moon all the way to Africa.

# ·III·

I was back to my novel full force in January, hoping to get two hundred pages under my belt before I had to take off for New York, there to confer with Whoopi and eyeball half a dozen manicurists in the second-story salons of Madison Avenue. Meanwhile the Writers Guild had started to convulse in anticipation of contract negotiations. I began attending meetings of a splinter group that came to be known as the Union Blues, its avowed purpose being to prevent a strike like the one that derailed the careers of thousands of working writers back in '81. John Allison called to say his musical was on, and I purred with understanding. Spring would be fine for the play, and until then I wanted nothing more than a chance to get my nymph and my curmudgeon to the end of Chapter V.

Late spring, it would have to be, for Roger and I had decided we were going to Egypt in April. Roger had sailed the Nile in '65 on a peasant ferry, all the way to Khartoum and on to the source of the Blue. He didn't really need another look at the Valley of the Kings, but he knew how eager I was to comb back beyond the Minoans to the pharaonic reaches of antiquity. Besides, we would be going the air-conditioned route, the Sheraton version of Cleopatra's barge. One way or the other, we figured to side-trip to Jerusalem to see his aunt and uncle. I promised Rog, who was usually the detail man when we traveled, that I'd take care of everything. He was too busy at work, and juggling all the minutiae of an apartment house we owned with his brother Sheldon.

Roger's parents were in town for a couple of weeks, on their way to the annual winter soak in Palm Springs, and while in L.A. they stayed with his brother in the upper strata of Bel-Air. We had dinner with Al and Bernice several times and one night packed them off to see *Holiday* because we'd enjoyed it so much ourselves. Roger told them how he had chortled over the hero, who meant to retire in his twenties and work later. That fierce decision to split for Paris at nineteen had occasioned a good deal of anxiety in his parents. It was heartening to see them all able to joke about it now.

Al and Bernice: my in-laws. They proved to be so heroic and so unflinching on the front lines that it's hard to recall when they were just the parents, benign in twilight. Al used to own a restaurant in Chicago, the H & H—downtown, streetside, real food, real people—all reflecting the Damon Runyon burnish of its proprietor. Owing to a bolt of angina, he'd been forced to retire in his mid-fifties, but he'd managed to parlay that brush with death into a long and vigorous retirement. Roger and I had flown in to Chicago to celebrate his seventy-fifth birthday in July of '83—a day after it was announced that a retrovirus had been isolated as agent of the AIDS infection.

Bernice had been a dancer as a young woman, a showgirl whose mother had sewed her costumes, and she still moved with a dancer's tensile strength. She'd never been hooked like *The Red Shoes*, though. Having given herself five years to make it, bound and determined not to stay too long at the tinsel fair, she gladly gave up her dancing to get married. Yet Bernice was surely the source of Roger's passion for high art. She painted, she read, went to concerts, didn't miss a beat at the Art Institute, and adored all manner of dance, from the clubs where she'd worked herself to the princess arc of ballet.

On January 20, 1985, Roger took her to the Joffrey on one of our season tickets, while I ate standing up in Sheldon's kitchen with Al. Bernice and Roger came home exalted, and we sat around having ice cream. Roger was fine. How is it I remember those moments sharp as a Kodachrome and see him perfectly healthy when I know now it wasn't so? I've got two pictures from January that spook me so much I can't even look at them now. One is Sunday the sixth, a birthday party for a relative once removed, in the parking garage of an office building, with every guest in a plastic hard hat coyly atop evening clothes. Roger looks ashen and drawn, though in truth I look pretty beat myself. Cesar had left only a week before. In the other photo, at the *Forty Deuce* cast party a week later, he's grinning and looks more himself, but I can see those shed three pounds in his face. Roger was built no-nonsense solid, five feet nine, 143 on the button, skin quick to take color, and never looked his age. In the cast-party picture he's older—not old, just older than I remember.

Roger must have mentioned to his brother that he'd lost some weight, because I recall Sheldon telling him one night to eat the fattening stuff, rattling off a merry list: potatoes, avocados, sour cream. Was it all dismissed like a joke because we were still in the age of lean-is-in? After all, there were body-mad men at The Sports Connection who would have paid equal weight in gold to lose three pounds. Not for very much longer, though. Within six months, lean—let alone thin—would become synonymous with the flashing amber of AIDS. In Africa they call AIDS the "slim disease." And even the compulsions of vanity don't hold up to fear. Thus in a year you would start seeing men at the gym who had chiseled themselves like Phidias now suddenly running to fat, the empty pounds accumulating in the waist and buttocks, evidence of the late-night binges on Oreos and Ring Dings that had replaced the faster food of bathhouse sex.

I'm not saying Sheldon's caloric list was a form of denial, not at that point, but it was very difficult to understand what Roger's brother thought about AIDS. In Chicago at Al's birthday in '83, I remember him saying he'd talked with a doctor who predicted they'd have a cure within a year. I took comfort in that for a while. Now, at the beginning of '85, he made it clear he didn't want to talk about Cesar. This mattered to me because Sheldon was openly gay himself, one of the few older men of the tribe I could talk to, who had seen the whole history of the movement, from the closets tight as Anne Frank's hideout to the broad daylight of liberation. I wished he didn't keep changing the subject whenever I mentioned AIDS.

He was a power broker who lived on a hill of money, who had made a huge reputation as a lawyer in the gay community when there *was* no community to speak of. Meanwhile he bought up bungalows and apartments in West Hollywood like Monopoly, and in the intervening years the mortgages were all paid off and the rents quadrupled. In the last few years he had opened Trumps, a watering hole on Melrose in which "foodies" gathered deliriously to graze, an overnight sensation, and had become chairman of the board of the Bank of Los Angeles. This last, in a former Rexall Drugs at the San Vicente corner of the Boys' Town strip, had been postmodernized within an inch of its life. Charlie Milhaupt called it the Tallulah Bank.

Many of those in power considered Sheldon the most important—or at least the most visible—gay man in California politics. He was close to Governor Brown and had addressed the Democratic Convention in 1980. Yet his constituency in the gay community was a curious one, since he also owned the trendiest bathhouse in southern California, wildly popular with the gym and disco crowd and noted for barring the door to men who were not "hot" enough. Sheldon's world was rife with small-town Adonises who came to L.A. for the good life. Roger and I always got rather tongue-tied around these men, feeling ourselves far too cerebral for the disco beat that roared through Sheldon's house. The two brothers lived in very different lanes of the freeway.

Yet it was a matter of some crisis in the family that both of Al's sons had turned out gay. Sheldon didn't come out to his father till '79, aged forty-eight, when he was made the centerpiece honoree at the first Center dinner. In saying yes to the testimonial he was already so far out that he had to tell his family. Of course they knew already, all of them, but knowing is very different from talking about it.

In '77 Roger and I had been living together in Boston for two years, and my parents had welcomed him into the family with pretty open arms. But nobody ever said the word *gay*. My own point of no return occurred when my first novel was about to be published. I slipped a copy of the bound galleys to my parents, figuring they'd better know what was coming before the book hit the stores. They reacted predictably, I suppose—telling me they would have to sell their house and leave town, that I'd never hold a job again with this infamy to my name, and besides, my mother had the idea I wasn't really gay. Roger was just a phase, I guess.

It's just an unavoidable mess, this coming-out business, and there don't appear to be any shortcuts through the emotions, though we try to make it easier for those who come after. Someday the process will be more human, perhaps, because we are open forever now, and people can't hate their children or themselves for that long.

When Roger finally sat his father down to tell him he was gay, we had been living together for six years, so obviously happy that to everyone around us we were a hyphenate, Roger-and-Paul, such a unified field had we become. Yet despite Sheldon's status as a gay leader, despite Roger's depth of feeling and his grown-up marriage, the double knowledge had thrown their father into a terrible depression. Roger and Sheldon were half-brothers, Sheldon's mother having died a generation ago. When Al went home to Chicago and told his doctor the news was twice as bad as he'd thought—*two* sons now—the enlightened doctor rolled his eyes and made it plain that Al must have done something very wrong. Thanks, Doc. For the next year and a half Al couldn't look me in the face, couldn't speak my name or enter our house. We got through it and were all very close now, as close as Roger had come to be to my family, but that doesn't mean that getting there was fun.

On January 22, Tuesday, Joel left a message on the machine: "Leo has AIDS." I called him back and learned that Leo was going into the hospital for tests. I remember getting very precise with Joel: There had been no diagnosis yet, it was still just speculation, in any case it was *pre*. I made the same pedantic distinction to Leo, as if the nightmare could be outsmarted with hairline distinctions, as if Leo could really care, feeling as wrung out as he did.

On Thursday the twenty-fourth we had dinner in Beverly Hills with a motley cross-section of the extended family—Sheldon's ex-lover and *his* current lover, various in-laws and out-laws, maybe fifteen altogether. They were talking that night about a miracle drug that had just been announced, guaranteed to curb baldness; no, actually grow hair. One receding fellow knew the name: minoxidil. He and Roger laughed that it had come along none too soon. I'm sure it was that night Bernice and I were talking about Cesar. She shook her head with compassion at some indignity he was suffering, and I shifted gears quickly and spoke with rigorous optimism.

"Well, of course, you have to have hope," she agreed, and though she meant it sincerely there was a gaping hole over to one side of the remark. For a moment I seemed to look over the edge into nothing.

I reached page 200 of the novel the same day as the dinner of the baldness cure. I don't remember anymore what happens right then in the book, but at the time I was thrilled at the rightness of it and torn about putting it aside to write manicure jokes. Yet all my notes from a session with Sam two days later are about mourning, what to feel about all the wasted time of your life, the wrongheaded decisions. I guess this must've sprung from some tortured version of *What if it all got taken away*? Mourning, said Sam, was a form of self-compassion. Looking back with sadness and asking why was the proof that one had grown. Don't just mourn; celebrate movement forward. As to how to get off the dime of all those squandered hours, the bottom line was this:

> *I know that when I was in Greece I felt* X *and I dreamed* Y *and it felt good. So onward.*

I had lunch with George Browning, an old friend who managed the mad American Express office in Beverly Center, the mall-in-the-sky at the southern tip of West Hollywood. George was delighted to sketch us a proper Nile itinerary, with a possible side to Jerusalem. But I had a new idea—I can almost hear my wheels spinning even now. We could do a week on the Nile, then fly home via Greece for a few days in the Peloponnesus, since I had to see Olympia now that I'd seen the Games. And *then* we could go to Israel in October, by which time our bonus mileage would be astronomical. We could do it all first cabin, coming home via Rome so I could spend my fortieth birthday perhaps in Capri, gazing out to sea from Tiberius's villa.

I lay all this out in its full hummingbird intensity because it was our last grope for the whole world, and none of it would happen. We had already had all of the world we were going to get. But you can't fault a guy for waving his ticket, even as the gate swings shut.

At 2 A.M. on Sunday, the third of February, I finished the last lined page of my bound notebook, the type I use for a journal. I don't appear in a millennial mood as I close the volume. The sheer casualness of it shows how much recovering one does from shocks like Cesar's visit or Leo's diagnosis. I still had a place to come home to, apart from all that:

> *The dog is sleeping in a curl beside me. . . . May this house be safe from tigers. . . . R & I both struggling with viruses, and we had a heaping bowl of oatmeal after the ballet.*

That is the first reference, right there, to the beginning of the end. But the twin flu is another sort of magic, homely as the oatmeal, for I felt safer that Roger and I were both under the weather. I knew deep down that all it was in me was a cold, ergo the same with him.

The day before, I'd had coffee with Carol Muske in the Valley. Carol is both a fellow poet and fellow iconoclast, with a laser wit and dead-center delivery. We always have a great romp, trading the silk scarves of literary gossip. Carol is married to actor David Dukes and mother to Annie, then two years old. She was as sentimental as I about Christmas, and I think she *did* string the popcorn and cranberries I drew the line at.

How it got started I don't know, but we'd spent nearly the whole time together talking about this tidal wave of doom we were feeling. It jarred the air around us like a siren din pitched just too high for human ears. My own horrors were all about AIDS now; Carol's were more in the deep cave of a mother's fear, where the only wakeless nightmare was any threat to Annie. Meanwhile Carol enjoyed my dispatches from the loop-the-loop of Hollywood, a writer's game at the exact polar remove from poetry. We respected one another's work without envy, since we weren't trying to be each other at all. In the next six months I didn't get saved a scrap of agony, but Carol proved to be the one who saved me as a writer.

On Friday, February 8, Roger had an appointment to show an apartment at one-thirty. I notice in his calendar for that week that he'd had to be there at noon on Wednesday too. There was always a pipe bursting somewhere. These gnatlike errands can prick a blood jet in me now because I remember how hard Roger worked on the house at Third and Detroit, how fastidious he was about details. There was a moment sometime in the previous month when he discovered he could buzz people into the building long-distance from his office in Century City, twenty minutes away. Till he knew that trick he'd always had to carve out an hour for Detroit matters.

I couldn't bear the details myself, being the sort who never would've gotten the mortgage or the taxes paid without a lawyer in residence. I expect he'd given up asking me to help out much; it wasn't worth the shrill of complaint from the resident word processor. This is where survivor's guilt and helplessness merge, because you start to think if only there were fewer errands to the apartment building, if only I'd picked up the Wednesday meeting, he wouldn't have gotten so run down. Maybe he would've been able to hold it back another two months. That way madness lies, I know, but you find yourself far down such paths in the woods before you know it. Then darkness falls and you're lost.

I was leaving for New York and the nail research on Sunday afternoon, so I had a torrent of errands of my own, from taking the dog for his shots to dropping by my doctor's for a dose of hepatitis vaccine. There was some vague theory around this time that hepatitis immunity was a line of defense against AIDS. People I knew still believed you had to have hepatitis first. As it happened, however, I'd already been through the six-month course of the shots and fallen in the one percent for whom the vaccine didn't take.

When I started the second go-round, I asked my doctor if presence of the AIDS virus couldn't have run interference and blocked the hep vaccine. Not too patiently, he explained for the nth time that I showed no sign of AIDS infection. He pulled out a learned journal and read the now-familiar catechism of pre-AIDS symptoms. You had to have at least two of these for at least two months to be *pre*, he said. He was growing very weary of this particular strain of somatic whine.

On Saturday night we went to a screening of Rob Reiner's *The Sure Thing*, because a swell lady of our acquaintance, Lindsay Doran, had been executive in charge of the production. The movie was half an inch deep but rather endearing, and in any case we were there to celebrate with Lindsay, who'd just landed a plum job at Paramount. There was a milling reception after the screening, which the three of us ducked to go eat at the Ritz Café, the Cajun beachhead in West L.A. As we were leaving, I bumped into a pair of writers who snarled with outrage at Reiner's little movie, attacking it with poisonous overkill. "It is not enough to succeed," as Gore Vidal first told me about writers in Hollywood. "All others must fail." Roger and I rolled our eyes as we left, relieved to have gotten past *that* stage at least. We laughed ourselves silly at the Ritz. Lindsay was the only Hollywood big shot we knew who had a hair-trigger aversion to taking herself too seriously. In this she and Roger were very much alike.

That was the last unravaged time. If Roger was feeling bad that night he didn't say, or maybe I'd just dismissed it as the tail of the cold I'd barely shaken myself. Sunday he was feverish and fluish, and I told him to relax by the pool and rest all day. I packed; I went to the gym for twenty minutes, pacing myself for a four o'clock flight. Midafternoon, Roger was in bed, yet his fever was barely ninety-nine. He didn't really have to take me to the airport, but he wasn't the type to spring the twenty-five for a needless cab. We were driving down Sunset, me at the wheel, when he suddenly said, in a frail voice with a quiver of tears: "Don't go."

I looked at him, startled. "I'll be back in four days. You'll be all right." As tender and reassuring as I could be, because I knew how unanchored we were when either of us was away. We'd always agreed it was harder on the one who stayed home. He fretted about feeling weak and awful, and I told him he'd be fine. I had a knack for soothing and calming him down, de-escalating crises. He did the same for me. Of course I would've canceled the trip if he'd pushed, but he didn't. I was sure he only needed to spill the low-grade misery and be stroked a little.

I arrived in New York about midnight and went straight to Craig Rowland's place, a Caligari railroad flat on Second Avenue in the Seventies. Craig was thirty-six, a free-lance journalist who lived on a wing and a prayer like all free-lance types. I'd met him the same night I met Roger. We'd been pals in Boston, then lost touch when Craig relocated to Houston and I to L.A., but had recemented our friendship during the two years he'd been in New York. Two minutes after I dropped my bags on the floor of his apartment, I showed Craig a sore on the shaft of my penis that I'd noticed over the weekend. "That's not AIDS, is it?"

He examined me closely and shook his head. "No, it doesn't look like that."

I immediately got in the bathtub, which happened to be in his kitchen, so I could soak it. Meanwhile we were talking about fifteen things at once. Craig had hung up the phone from Houston just as I'd arrived; a friend was calling to tell him they'd pulled the plug that morning on a young man who ten days before had seemed as well as you or I. Suddenly Craig pulled back the sleeve of his flannel shirt and showed me his arm. "What about this?" he asked. I looked at a small red spot above his wrist, slightly raised, barely a quarter-inch across. "No way," I said. "They're never raised."

I was wrong.

But we didn't know any better that night, and we joked for another half hour before he went to bed. I cozied up in the loft in the back bedroom and called Roger. We had a marvelous chat, full of our private ironies and shorthand. He was feeling a bit better and more rested, and promised he'd take care till I got back on Thursday. Craig recalls my exhilaration next morning, telling him what a good talk I'd had with Roger.

At dinner that evening I ate a single mussel out of a colorless bouillabaisse, and within four hours I was violently sick, groaning all night in bed without the wherewithal to vomit. I was bent double with cramps as I walked in the pouring rain on Tuesday to the Russian Tea Room to have lunch with Whoopi. I can't imagine what she remembers of that occasion, if anything, though it must've dismayed her considerably to think this humorless man sipping broth and Coca-Cola was meant to be her breakthrough into feature comedy.

I'd talked to Roger midnausea the night before, and he was complaining again about feeling awful, but for once we had to call it a draw. Tuesday night I was stronger after spending most of the day in bed. I could hear how rattled Roger was when I called him late: he wasn't just sick, he was worried. But if it was AIDS worry it was still unconscious. He never thought he'd get it; he said so in the hospital. It was harder and harder to soothe him from three thousand miles away, and now I was near frantic to finish up my manicure research and get back to L.A.

Wednesday night I had dinner in the Village with Star Black, my oldest friend. Star was a photojournalist who was also a closet poet. One Sunday in 1968 we'd driven out to Newton, Massachusetts, and sat in the car in front of Anne Sexton's house, I'm still not sure why: for purposes of osmosis, perhaps. It was always a great pleasure for me to encourage Star's writing, but that night in '85 all I could talk about was my fear for Roger: *What if he has AIDS?* I expressed it openly for the first time. Star's response was as adamant as it was instant: Impossible. We'd been safe too long; neither of us had been really sick; I was overreacting to all those other cases. I don't think she succeeded in calming me down. But this notion that we had somehow squeaked in under the wire to a sort of viral demilitarized zone, while it didn't comfort me then, became one of the totems I clung to through the next unyielding weeks.

I left New York on the fourteenth, badly shaken. All the dazzle and energy I connected with the city—cinematography by Gordon Willis—seemed utterly dissipated now. On the way to the airport a speeding car in the next lane hit a dog, who went yowling off up a side street. On the plane I sank into my first-class studio-paid seat and talked to no one, I who usually acted as unpaid social director on a flight. I buried myself in a biography of E. M. Forster.

The plane was fitted out with a novelty item, an air-to-ground pay phone. I called Roger to wish him Happy Valentine's over Kansas. He was home in bed, too distracted by fluish aches and pains to appreciate the call. We were to meet that evening—I was going direct from LAX—at the Variety Arts Theater, where Roger was sponsoring a table at a benefit for Room For Theatre. I figured we'd play a little ship-to-shore, but the connection sucked, and besides, he wasn't in the mood. I went back to my seat and forced myself to finish the Forster. I wrote out my notes from the visits to all the haute salons. The in-flight movie was *Garbo Talks*, but I didn't take the earphones. I needed to keep busier than that. Yet there was a random moment when I looked up to think and stared unexpectedly at the silent screen, and Anne Bancroft was lying in bed in a hospital, dying. I started to heave with sobs. The stewardess gently skirted by me with the liqueurs, as if I were on my way to a funeral.

I shlepped my bags from the airport to the Variety Arts, and went limp with relief when Roger came in, beaming, with Alfred Sole. A look was all we needed by way of anchorage. The evening was structured as a musical revue, and I remember sharing a grin with Rog as a high-camp comedienne did a flaming send-up of the benefit circuit. Even so I was feeling annoyed that he'd been stiffed to pull together yet another charity affair. We already did this sort of thing twice a year for the brothers and sisters. I felt a surge of protectiveness, as if people were taking advantage of Roger's benign nature. He couldn't say no to a good cause. Yet he seemed in terrific form that night, bantering with his fellow board members, in tune with the collective goodwill as usual. Unlike me, who soured easily these days, on the edge of burnout.

Next day he came home early from the office and went to bed. It was then I told him he really had to call Dr. Cope at UCLA. Not to panic, I quickly added as he winced apprehensively, but he could be harboring some sort of low-grade walking pneumonia that needed antibiotics. He agreed with a certain relief, comforted just to be talking about it as a concrete thing with beginning and end.

I might note here, with an odd dispassion, that while I was back east the carpets were cleaned, and all through the house were little white squares of Styrofoam under the feet of the furniture. Two years later there are still tables and chests with the Styrofoam crumbling beneath. As I say, it's the details that get away from me.

I cooked stevedore meals and pampered Rog all weekend. "Everyone's got the flu, it's all over the place, everyone says," I wrote two weeks later, grasping at the straws of the flu tautology. Yet from the moment I said the "A" word out loud to Star in the dive on Irving Place, I had crossed a line. I spoke openly now of this marrow terror of sickness—not to Rog, I didn't want to upset him, but to certain gay friends. I know now I uttered the word as a sort of reverse hex, as if by daring to speak I would neutralize its power. Being scared is not the same as being convinced. Fear still has room to maneuver, and every wave of its energy goes into pushing the terrible thing away, like the ocean leaving a body on the sand.

My journal gets very spotty here, with only a single detailed entry for the whole sea change of the next six weeks. As if the record itself didn't know how to stop taking its cue from Mrs. Woolf and learn to be *The Plague Year*. I know the refrain of the next month, from every side, was constant: *It's not, it can't be*. Roger was the last man anyone thought would get it, just as Leo had been in his circle. To the gay men around us, admitting the shadow had fallen on Roger was to unleash a wild surmise naked as the pandemic across the belly of Africa.

My memory of those weeks, back and forth to UCLA, is mostly shell-shock fragments. I can't even put them in chronological order, let alone weigh them. I know Roger spent three days in bed, then tried one at the office, only to wilt and crash with another fever. That would be February 19, the day the water pipe burst in the bathroom off the guest room and another burst on Detroit Street, so we had plumbers slogging in and out, distracting us with the chaos of banality. Thursday Roger put in a full day at work, came home with 101.6 and logged yet another weekend in bed. He would seem to get stronger if he laid low for a day or two—he wasn't getting worse, just not getting better. People would drop by to visit, full of statistics about the alphabet of influenza coursing through the city like an ill wind from the East. How much denial was everyone practicing? Enough to power Chernobyl, but nobody did it consciously; that's why it's called denial.

Then Monday the twenty-fifth, Craig called from New York. He was reeling from a session with a doctor who had diagnosed KS right in the office, after the briefest eyeball of that minor spot on his forearm. Since the biopsy would take a couple of weeks to be certain—laboratory backup, one of the essential elements of plague—Craig was on his way to Houston the next day. He had high-level connections in the medical world there, having been founding VP of the Houston AIDS Committee. He promised he'd call as soon as he knew one way or the other. He knew already, but there is the matter of the formal confirmation. One spreads the shock as best one can over several days. I wasn't much help from my end, having suddenly developed amnesia in the positive-attitude department.

I was also running out of friends who weren't sick. Now I began to hyperventilate with panic and claustrophobia, the stakes seeming to double every time the phone rang. What morning was it that I first woke up suspended in that instant before a car wreck? The hysteria first to last was much more acute in me than in Roger. It's with a certain awe I look back and see how balanced and focused he stayed, even as he gathered and husbanded his strength, patiently trying to get back to work. It's not just that he wasn't a complainer, or that his attitude was stoic. That would come. It was rather that he took refuge now in his temperate nature, a capacity for quietness that began as instinct and ended as character.

There's a complicated Greek idea that the Greeks pared down to a single word: *sophrosyné*. R. W. Livingstone, the Oxford don who translated the Plato we read a far summer later, describes the force field of the word with eloquent high-mindedness. *Sophrosyné*, he says,

> *stretches out and tends to become the whole of virtue, an inner harmony of the soul, a reasonableness which reveals itself in every action and attitude. In war-time it vanishes almost entirely—especially among civilians. It is, in the literal meaning of the Greek word, 'soundness of mind'. Restraint is of its essence, but is felt not as restraint . . . but as that natural service to right reason which is perfect freedom.*

He wrote that in 1938, at the end of another world. There isn't a nuance of it that isn't full of Roger, all the time I knew him but especially through his illness. If I idealize him out of proportion in saying so, well, beware the storyteller then. I'm with Livingstone. You're not supposed to have to be a hero to embody such a vastness. The whole of Greece used to work toward it. If it seems rare and outsize now, that only says more about an age resolved to face the millennium without it.

More to the point, if Roger had great patience, I have none. Here at the pitch of emergency I can only lay out the fragments of what seared my frantic heart. I am the weather, Roger is the climate, and they are not always the same. Yet the careening of those next few weeks, fitting in visits to UCLA, more and more in tandem, is the story of a kind of bond that the growing oral history of AIDS records again and again. Whatever happened to Roger happened to me, and my numb strength was a crutch for all his frailty. It didn't feel like strength to me, or it was strength without qualities, pure raw force. Yet it took up the slack for Rog, and we somehow always got where we needed to be. In a way, I am only saying that I loved him—better than myself, no question of it—but increasingly every day that love became the only untouched shade in the dawning fireball. What Tillich calls God, the ground of being.

Roger's blood was drawn fifteen different ways, but we had no test for antibodies yet, so none of the numbers led anywhere. Still there was no perceptible cough, and the general malaise and zigzag fever weren't in themselves conclusive, could still be that phantom flu, shimmering now like an oasis. During one of his consultations, Roger came out to the waiting room and said Dr. Cope wanted to meet me. The feeling was mutual.

As soon as we sat down with Dennis Cope I silently took back every idiot pun I'd ever made about his name. He's a bear of a man, seized with concentration yet extraordinarily mild by way of affect. Speaks carefully but not guardedly, and never to cover his ass. We were three ways blessed: that he was brilliant, that his reputation gave him power, and that Roger had been his private patient for five years going in. Dennis Cope and Roger already had each other's measure before they ever engaged in this battle together. Modest to a fault, incidentally; doesn't even hear praise. And not once in twenty months did he not have time.

He was perplexed the day I met him, but proceeded methodically and threw up no red flag that I could see. He said we had to keep probing these tentative symptoms, but no, whatever it was didn't present like AIDS at all. For one thing, Roger wasn't sick enough. If that sounds naive two years later, I have to remember the syndrome was defined then only by its direst fulminations—gasping on a respirator, lesions head to toe like shrapnel. Roger didn't exhibit the requisite pair of *pre* signs, or not sufficiently to chart a downward curve. Maybe a doctor in New York would've been more grimly fatalistic, like the dude who flattened Craig in a matter of seconds. Maybe Cope was growing more worried and chose to protect us. But so far he still appeared to subscribe to the stubbornness of bugs, just like all our dutiful friends.

How far was that? On March 1 he told us the chest x-ray looked clear, except for a shadow that was probably the pulmonary artery, but he was playing safe and ordering a CAT scan to make sure it wasn't a lymph node. Roger and I had lunch that day at the hospital cafeteria, in the prison-yard court on plastic chairs under a lowering sky. Roger said how glad he was I was there. My sentiments exactly: as long as we stood our ground together we could thread our way through this maze a step at a time, Buddha's way to the top of the mountain.

A few days later when Roger went in to see Cope, I ran out to the corridor and called a friend, to grill him as to his own bout with "regular" pneumonia the previous winter. There had been a suspended day or two back then, as we all waited uncomfortably for the man's results, and then the tests proved negative for AIDS, and we all went back to life. Now I gripped the phone white-knuckled, hammering symptoms out of him. One by one I compared them to Rog, pinning my case on that regular brand of infection. When I strode back to the waiting room, Roger was sitting there stunned, and he stumbled out into the hall as if my five minutes away had nearly let him drown. He sagged beside the water fountain and spoke in a kind of bewildered shock: "He says it could be TB."

Then he started to cry, and the burst of tears sent one of his contact lenses awry. So instead of holding him I had to cup my hands under his eye while he worked the lens back in, swallowing the scald of tears. That specific helpless moment, the soft disk swimming out onto his cheek, stuck with me like a pivot of agony. A year and a half later I'd still be trying to explain to Rog, when the talk came round to the horror, how in that noon moment I died inside. As if I would not live in a world where my friend could be in pain like this. I don't remember what happened then, if we had another test or were given leave to go home, but something had cracked that would never knit again.

As to how we so tenaciously continued to deny it, I offer one morning when I let Roger off at the main entrance to the medical center. He was going to the eighth-floor pulmonary unit to have an arterial blood-gas test. I parked the Jag in the underground garage and was lurching across the plaza when Rand Schrader happened to come out of Jules Stein Eye Institute. As soon as we saw each other I began to weep, and Rand waited till I was calm and walked me all the way to where Roger was, through a labyrinth of corridors. He says now it was obvious that Roger was very ill, and the test in question extreme. They sink a needle into the artery at the wrist and sip out a vial of deep blood.

Rand doesn't remember, but I do, his telling me as we walked about an acquaintance in San Francisco who was out of the hospital after a bout of pneumocystis, back to work and fit to travel. This was supposed to be encouraging. It was, in fact, so deep had the needle sunk now. My panic had evolved to the more encompassing fear that Roger was dying. If it kept getting worse, Death would start sniffing around, no matter how incomplete the diagnosis. Rand stayed through the test, then waited with Roger outside while I brought the car around. Yet he says by the time he got home he'd buried the whole episode. He bought our story two weeks later that Roger's pneumonia was normal as a football jock. He didn't want to know yet, and I don't blame him. Once you know, it's all over.

A couple of days afterward we were eating a glazed breakfast before going off to UCLA. As I cleared the table—things in order if not life—Roger looked up at me and said: "It's just the two of us."

"I know," I replied, though of course we weren't alone. Al and Bernice had left for the desert on February 1, but they were calling in regularly to check up on us, clearly very anxious. Roger had been looking forward all winter to his sister Jaimee's arrival in Palm Springs with Michael and the kids, yearning to put this thing behind him so he could go play uncle. We had two doctor friends we ran the numbers by, and the phone was constant with friends' concern.

All the same, it was just the two of us lining up as the tests grew more harrowing, the corridors at UCLA more like a separate equal world every day. Forgive us the feeling now and then that the woods had closed behind us. In the most visceral way, with a taste like a ball of blood in our mouths, it seemed that life itself was pulling in like a tortoise. Inside its armor crouched the "group of two" that Freud calls a marriage. Not career, not the past, the waste of errands or the state of the planet. Just us.

I was taking Roger's temperature every couple of hours now, shaking down the thermometer till I had a twinge like tennis elbow. One crazed afternoon I accidentally broke the thermometer against a door and fell to my knees keening, trying to pick up the shards as the mercury beaded into the jade-green carpet. I was cooking in twice a day, shopping at Irvine Ranch, bewildered by the sunny vigor of what Randall Jarrell calls the "basketed, identical/Food-gathering flocks," with their nine-dollar purple peppers and the BMWs in the parking lot. Already I went about in public as if I were on the moon. I had to ventriloquize my way through various meetings at the studio and with Alfred. I began calling my brother late at night in Pennsylvania, needing his constant reassurance after Roger went to bed.

"Paul, it's not AIDS," Bob said over and over, though he knew the tragic randomness of things far better than I. He was born with spina bifida, and had been in a wheelchair all his adult life. Six years younger than I, he spent months at a time in hospitals as a child. I have a vivid memory of visiting him with my parents in Springfield at the Shriners Hospital. Because I was under sixteen, I had to stand in the bushes and peer in a window and wave at him, lost in a ward of suffering. He's one indomitable character, my brother—an accountant's accountant and a teacher, married to his high school sweetheart, Brenda. Throughout the skirmish over my novel, Bob had been the buffer zone between me and my family, the one who understood being gay, who understood being a writer. In the summer of '83 a drunk plowed into his car head-on and put him in the hospital for a month. Oh, had he been there.

All through the week of March 4, Roger's calendar is full of precise notations: 2.0 Godino will, 0.3 Scott Redman. The hours of a lawyer are broken down into tenths, and he kept the record in his calendar because he was home in bed, but he wouldn't stop working. The blood-gas results proved to be in the normal range, which was a relief, yet there was clearly some kind of infection in the lung. The issue at week's end was whether or not that infection was "interstitial." Pneumocystis carinii—the deadly AIDS pneumonia, so-called PCP—is an interstitial infection, which means it invades the interstices between the lung sacs. A battery of x-rays seemed to indicate no interstitial involvement, and this was taken to be good news, especially by our doctor friends, Joe Perloff and Dell Steadman. Joe is a research cardiologist, Dell an eye surgeon. We were pinning them down for opinions in matters that weren't their field, but they were generous here as they would be throughout. Once I heard the interstices were clear, I tossed the pneumocystis file away. You become very primordial about data. What you need you eat whole, like a python consuming a rat. What doesn't apply right here right now is moontalk.

Thursday or Friday a letter arrived from Craig. The doctor in Houston had confirmed the diagnosis: AIDS by reason of KS, no treatment at this time. Craig was writing to eight or ten friends to break the news, but otherwise he wanted to keep it private. He would widen the circle at his own pace. It couldn't have been more lucid or dignified, and I read it to Rog like a bulletin off the Kafka wire service. *When is enough*, I kept thinking, as if every tragedy mounting up would finally satisfy some savage god.

Despite the positive sign on the interstitial front, Roger still wasn't getting any better. Still not worse, but Cope decided it would only be prudent to have Roger come in for a bronchoscopy, in which a flexible tube is inserted in the lung for a specimen of tissue. The bronc has become such a fact of all our lives now, it's hard to recall there was a time I'd never heard of it. Joe Perloff promised the test was remarkably negotiable, though I recalled Joel saying the doctor had managed to puncture Leo's lung.

Roger would have to go into the hospital overnight to have it. Neither of us had spent so much as a day in that nether place, not in our whole ten years together. Till then I affected to feel rather phobic about the whole idea. The previous fall, when Kathy Hendrix had been in for surgery, I told her over lunch a week later that I hadn't been up to visit because I wasn't good at hospitals. I think I still clung to the trauma of the past, pressed against the Shriners window, like a shield. I would learn now to put such bullshit behind me very fast, and afterwards would feel a kind of nuclear contempt for those who practiced it anywhere in Roger's orbit.

Over the weekend before he went in, we just hunkered down. I'd finished reading Forster and turned to *The Golden Bough*, as preparation for the powers of Egypt. Every night I would pore through Frazer's laundry-list account of magic and fear and atavism. I kept beside me a folder of Nile cruises, which I would scribble with lofty problems: Did it matter if we booked port or starboard? Mostly what I was doing was repeating the interstitial news like a mantra, over and over, to drown out the week's other blip of evidence. Roger had failed the scratch test for mumps—had shown no red or blistering when the patch was removed from his arm—and this was considered a crude sounding of weakened immune function. If the choice was either impaired immunity or an unreliable test, I was for betting the farm against the test.

Roger was comfortable resting in bed, still no cough to speak of, animated with everyone who visited. Saturday night he convinced me to go out for dinner with the Perloffs and the Rankaitis/Flicks. These were the couples we saw most often, who gathered around them the most stimulating people, mind over Mammon. Marjorie Perloff was then at the University of Southern California, an encyclopedic and inexhaustible literary critic who knew every cusp of modernism backwards. Robbert Flick and Susan Rankaitis both make photographs, but the camera is merely the common denominator here. The light these two work by is opposite as sun and moon. Marjorie and Joe, Susan and Robbert, Roger and I—we had constituted an inner circle for many years.

At the restaurant I made Joe explain the interstitial data all over again, and he tried to ease my mind about Tuesday morning's test. Finally I lightened up enough to eat. Susan says she never suspected Roger had AIDS till I told her seven months later, so Joe presumably succeeded in reassuring somebody. When I got home, however, I found Rog sitting in the study coughing, and looking more drained, worn out and lost than he had all month. He was so glad to see me and be taken care of. At such a moment you move like an avalanche to oblige, for all the reasons of love but also just to keep busy. It was going to be fine, he'd be home by Tuesday afternoon, and after that there were no more tests. Then he would have to get better, I thought, as I kneaded his shoulders and curled him to sleep. I read *The Golden Bough* till 4 A.M.

Al and Bernice had decided to drive in from the desert, even though we told them it wasn't necessary. They agreed to hold off till Monday and just stay overnight till we got the results of the bronc. On Sunday evening I fixed the two of us as cheery a supper as I could muster. Roger's appetite hadn't suffered, at least. But in the middle of the meal he excused himself to go to the bathroom, where he had a twinge of diarrhea. This didn't prove to lead to anything ominous on its own, and in fact through all his sickness Roger didn't have to deal with much intestinal static. But as I sat at the dining room table waiting for him to come back, the food like ashes on my tongue, I had a sudden vision of what a flimsy wall we'd been building the last few weeks, brick by brick.

He seemed so weak and overwhelmed by then, and the hardest thing to watch him lose in the early days was the spring in his step. He'd always had a quickness about him, a vigorous enthusiasm that I can still see in picture after picture out of the past, like a great store of potential energy. The wellspring of it wasn't athletic; it flowed from a joy of life. In the steepening decline of the previous months he'd lost the physical edge of that delight—lost it for good. Though he had reservoirs of deeper and sweeter tones to compensate, I missed the boyish energy most. Perhaps because mine went with it.

Sunday night, *Vertigo* happened to be on television. We'd both seen it decades ago but never since, especially not with the drift of learned exegesis that has developed around the fifties Hitchcock. We lay close in bed to watch it and were soon transported into its spiral subtext. The worry about tomorrow seeped away a little, or maybe it was just a relief to watch somebody else—in this case Jimmy Stewart—be torn apart by suppressed hysteria.

On the way to UCLA on Monday morning, driving along Sunset to the west side, Roger asked quietly: "What if it's really serious?"

Despite the positive talk all week—all month—and despite the fact that my last nickel was riding on denial. I don't know if I answered the right question, but I know my voice was steadier than I would've thought possible. Rog, I said, you have to understand how much everyone loves you. He had nobody out there even approaching enemy status; I'd never heard anyone say an unkind or quarrelsome word about him. The same could not be said of me, by a long shot. I gave a little encomium on his talent for friendship and loyalty, the idea being that everyone would be there for him if the going got rough. I'd learned this tactic of human grounding from Roger himself, who would always be saying as we drove away from a dinner party, about someone I hadn't even noticed, "Such a nice man. So unpretentious." Unless we were driving home from a migrainous Hollywood party, in which case he might grumble about some hustling producer or other: "Too noisy. So full of himself."

I can't really separate the March 11 check-in on the tenth floor of the medical center from a dozen others. Amateurs still at the system, I expect we appeared like two meek refugees, with the overnight bag and a briefcase full of work. The tenth floor at UCLA is called the Wilson Pavilion, all private rooms and food prepared to order, the carpeted veneer of a hotel corridor not quite masking the naked high-tech sick gear. There was a waiting room across from 1028, dedicated to Nat King Cole, where we plunked the parents down with their thousand-page potboilers. I stayed with Roger throughout the day, working in fits and starts on *The Manicurist*, as the interns came in and drew his blood.

We would both grow grimly accustomed to the first day of a hospitalization, with the interns sweeping in as if by revolving door, trying to look serious in spite of their comical youth, mad with backed-up things to do and racing like the White Rabbit. There would come a time when I would take over this phase, give the tedious history, answer the bald questions: Are you a homosexual? Are you or have you ever been an IV drug abuser? On March 11 I couldn't tell one intern from the next, intern from resident. I didn't realize that in a teaching hospital like UCLA every patient is one more unit to cover as they cram for the test of their budding careers. And here in the presence of a new disease, each kid doctor wanted an A. But remember, Roger was only supposed to be there overnight, so I held them all at arm's length and resisted differentiating.

Roger bore the process very well, and we seemed to be taking a proper stand of firmness in saying he was feeling not too bad. *Not sick enough, not sick enough*—I kept repeating Cope's phrase. It was still so, wasn't it? The pulmonary man came in to explain how the bronchoscopy worked. They would do it early in the morning, and we should probably have the results by noon. Home for lunch.

Dennis Cope was a welcome sight late in the day, because he at least knew who we were, and more to the point, the interns knew who he was. That is one of the shocking things about a hospital: its leveling of you to your body's weakest link. The Ph.D. in Comp Lit, the years in Paris, the wall of books—you do not wear these badges on your johnny gown. No wonder I was forever giving our résumés to doctors and nurses, as if to beg them to see us for real, see what happy lives we had left at the border, which waited still like a dog on the front stoop.

I must've gone out for dinner with Al and Bernice, and I must've been full of reassurance and interstitial data. All the blood work was normal so far, but I don't recall if an actual T-cell test was taken, or if we knew the results before the verdict. The T cells are a subset of the white blood count. Infection with the AIDS virus reverses the T-cell ratio, indicating an immune dysfunction. A test was available at that time, but it was still considered exotic and far down the line of inquiry. Today I know fifteen people who have their T cells tested every six or eight weeks.

I also wonder now, in a sort of stupor, how it was we had no plan whatsoever if the news was bad. We hadn't ever discussed who would know and who wouldn't, how we would euphemize, indeed if hiding was even feasible. In a way it was like the whole last year, when we never talked about dying because we were fighting so hard to stay alive. I understand that in theory it's good to have these matters out, to make one's lifeboat plans and release the sum of one's worldly goods. But we didn't seem able to do that and forge ahead at the same time. Warriors in pitched battle do not make their last will; they become it.

The final thing I remember from the night before was a visit from Michael Gottlieb, the immunologist who'd reported the first four cases of the disease to the CDC in the summer of 1981. Dr. Gottlieb was an intense man with darting foxlike eyes, who probably hadn't had ten minutes to catch his breath in the four years since he grasped the iceberg's tip. He spoke casually enough with us and said—I think I'm right about this, but who can sound the depths of my longing to hear a good omen?—it would all probably prove to be nothing. Then Roger asked him specifically: "Have you ever seen anyone with my symptoms who turned out to have it?"

"Yes," said Gottlieb.

My brother and Sam, my therapist, concur that they talked to me late that night and gave me the final pep talk: *It's not AIDS, it'll all be fine tomorrow, get some sleep*. Did they really think that? I ask them, and they both say they don't know anymore. They realize they were in shell shock then, to a lesser degree than I but with the same dazed sense of staring into headlights. How was I to know my very advisers were locked in a vertigo precipitous as my own?

I was over at UCLA on Tuesday morning before the pulmonary

team, and his parents and I gave Roger a bracing squeeze. I stayed with him till the doctor came in to administer the local anesthetic, and then I waited in the empty lounge with Al and Bernice, watching them as they dutifully read their books, refusing to leave my watch when the two of them went for coffee.

Altogether it took maybe twenty minutes. I was in Roger's room the second the team walked out. Roger was lying on his side, with an oxygen mask over his nose and mouth. They'd told us he would need it till the anesthetic wore off. What they hadn't said was that he would be coughing, almost without stopping and clearly in real discomfort. I patted him and talked a bit, but we really couldn't communicate. Even if this was a predictable reaction to the procedure and nothing more, the reality was jarring in the extreme. For this was the very cough we'd always said wasn't there. Could things have changed so fast? And who among my advisers would have me not worry now?

I sat in the red vinyl chair in the window corner and worked on another half-page of dialogue. If comedy's roots are pain, those must have been hilarious lines. Thankfully, Al and Bernice didn't come into the room; we were better off alone till we could talk again. I guess we had already arrived in that leveled place where nobody could follow, the only thing worse being the portal from which I would be barred myself, nineteen months and ten days later. I don't think there was any magic left in me as the clock ran down its final minutes. I was struck with a fit of the metaphysical bends, equal at least in hollowness to that bruised and hacking cough.

Finally it abated, and the oxygen mask came off. Roger was so debilitated from the trauma of the test that he lay back in an exhausted sleep. I don't know how much time went by. When the doctors came in—a pair of them, the intern and the pulmonary man—they stayed as close to each other as they could, like puppies. They stood at his bedside, for the new enlightenment demands that a doctor not deliver doom from the foot of the bed, looming like God. The intern spoke: "Mr. Horwitz, we have the results of the bronchoscopy. It does show evidence of pneumocystis in the lungs."

Was there a pause for the world to stop? There must have been, because I remember the crack of silence, Roger staring at the two men. Then he simply shut his eyes, and only I, who was the rest of him, could see how stricken was the stillness in his face.

"We'll begin treatment immediately with Bactrim. You'll need to be here in the hospital for fourteen to twenty-one days. Do you have any questions?"

Roger shook his head on the pillow. I wanted to kill these two ridiculous young men with the nerdy plastic pen shields in their white-coat pockets. "Could you please leave us alone," I said.

And they tweedled out, relieved to have it over with. I ran around the bed and clutched Roger's hand. "We'll fight it, darling, we'll beat it, I promise. I won't let you die." The sentiments merged as they tumbled out. This is the liturgy of bonding. Mostly we clung together, as if time still had the decency to stop when we were entwined. After all, the whole world was right here in this room. I don't think Roger said anything then. Neither of us cried. It begins in a country beyond tears. Once you have your arms around your friend with his terrible news, your eyes are too shut to cry.

The intern had never once said the word.

# ·IV·

The first thing we did was call Sheldon. He must have told us to, because I can't imagine why I wouldn't have gone right across the hall to break the news to Al and Bernice. Sheldon said not to do anything till he got there, and since nothing was what we felt like, it was an easy order to follow. Fifteen minutes later he walked in, vigorous and very calm. His attitude was so startling, so unlike our own dumb aftershock, that we both looked readily to him for direction. He was relentlessly upbeat. Not surprised about the results, and the only thing to do was get the infection taken care of and back to work. There could be no question of telling Al, because the shock might kill him. Why burden the parents anyway? In fact we couldn't tell anyone, for as soon as word got out, Roger would lose his practice. AIDS was that extreme a stigma. The losing of jobs was a foregone conclusion.

It was an insidiously airtight argument, and Roger and I agreed immediately. After all, I'd already promised Rog we were in this thing to win, and what better way of winning than to go on as if nothing had changed? I knew from Craig's example the week before, his lightning trip to Houston, that there were people out there who knew more than anyone told us. But I made it plain to Sheldon that certain ones would have to know. My brother, my therapist, the Perloffs—Joe had been following the case for two months. Sheldon resisted every name I mentioned, especially Rand Schrader and Richard Ide, as if gay men in particular couldn't be trusted not to gossip. It almost came to an argument, but I wasn't about to be budged, and we couldn't start bickering in front of Roger.

Having sealed the smallest circle he could get, Sheldon, brimming with optimism, sailed out. I still don't know why we allowed him so much power, except that he'd always assumed the role of head of the family. He made it seem so much easier to island ourselves with the secret, putting off the cold reality of all our loved ones' grief. I think Roger couldn't endure the thought of pity, he who was always so self-contained. For him at least, the secrecy didn't spring from shame, though I had a very bad case for a while. The shame of being different was rooted deeper in me than the fact of being gay: in the mesmerized faces of those who would stare at my brother on crutches, my brother too busy walking to notice. I noticed.

It was left to me to walk across the hall to Al and Bernice, still innocently reading, and tell them the good news: Not AIDS. Their burst of relief was nearly as hard to take as the horror a bare half hour before. I hated the lie right away, but there was nothing I could do about it now. I explained that the organism in the lung was bacterial and would respond nicely to Bactrim. The mere brand name would tip people off these days, but not then. Roger was still only the second person I knew after Leo to be struck with pneumocystis, and I thought I knew everything.

The day grew more and more surreal. There was a near-celebratory air around Roger's bed as his parents cheered him, giddy from the release of tension. At the same time we had to be sure that Dennis Cope made himself scarce for the next twenty-four hours, since he didn't feel he could lie bald-face to the parents. Incredibly, Roger's calendar records a telephone call with a client that afternoon. At least until we could get Al and Bernice off to Palm Springs, we were to go on as if everything was thumbs up.

The IV team came in and plugged into Roger's forearm—he had beautiful veins there, like a Renaissance bronze. The Bactrim drip was started. When Gottlieb the immunologist appeared to say he was sorry and offer his support, the first words out of Roger's mouth were: "What about Paul?"

His anxiety was palpable. Gottlieb reassured him with a ballpark figure: only ten to fifteen percent of the partners of AIDS patients had broken through to full-blown infection. *Yet*, he might have added. I looked at Roger with a pang of unworthiness, that I should be the main thing in his mind as he teetered on the cliff edge. Gottlieb turned to me. "You're a writer?" he asked with a skeptical air. "Why don't you write about this? Nobody else does."

For his part, Dennis Cope couldn't have been more supportive when we told him we chose to see the diagnosis not as a death sentence but rather as a life challenge. *I* made that announcement, and I know how it rang with a dare to be contradicted. In turn, Cope assured us that the infection had been caught at a very early stage and should eradicate without difficulty. Although people used to die fast from the initial bout of PCP, great progress had been made in zapping it. We were absolutely right, he said; the first glimmers of treatment for the underlying condition had begun to break in research circles. His confidence affirmed our own; his pledge to fight beside us was unswerving. No reason at all that Roger shouldn't get strong and go back to work.

I'd made an engagement weeks before to have dinner that night with an old friend from Yale, David McCarthy, now a cardiologist in Philadelphia. He was in town for a medical conference, and I held to the party line of normalcy and went ahead with the plan. Over dinner I told him everything had turned out fine at UCLA, but I've come to realize from his silence since that trip that he wasn't buying any of it. How could he, with me so dark and disconnected, foisting off all my AIDS insanity on Cesar and Leo—trying to explain what the vortex was like, even as I masked the cracks in our own caving house.

Roger and I agreed that he would deal with the parents on Wednesday morning and see them off to Palm Springs. I would stay as scarce as Dennis Cope till they were on the road. But Roger called in anguish the moment they left his room on an errand. "I can't stand this," he burst out. "I'm so sick of my father telling me how lucky I am!" I anchored him by narrowing down to the short term: Just get through the next hour and they'd be gone. By the time they returned two weeks later on the way to Chicago, he'd be on his feet and back to the office.

I still have such conflicted feelings about hiding the diagnosis. What's privacy and what's denial? How much is guilt and the lingering self-hatred of the closet? I've tried to fever-chart the stages as they evolved over the next six months, but the simplest way to put it is that Roger fought his way back to real life, and I fell completely apart. Till the end I was sure we'd made a mistake holding back so long from the full resources of everybody. I blamed Sheldon for a decision we colluded in with open eyes. Now I've done a total reverse: I'm glad we did it the way we did, that Roger had those six months more of his law practice. But what about the parents, so brave and loving once they knew? Did we really think Al would have a heart attack? That we would be punished through the pain of others?

The privacy issue surrounding AIDS engages vectors of the nightmare that make it different from every other medical crisis. I know half a dozen men who are dying right now, another dozen diagnosed, and everyone's being kept out. Of course the pose has worn very thin. We are all too aware and paranoid to be fooled by the regular brands of pneumonia. Not back then. The difference in just two years is exponential ground: two or three cases in every gay man's life are ten or fifteen now, and the clock keeps counting even when nobody knows or dares to ask.

I took on the lion's share of the calls among our friends, telling everybody it was all okay, then pulling back and shutting them out. I know I had to argue rigorously with Roger after the parents left to let me include at least one friend in L.A. I suggested Richard Ide as being less likely to be pressured for the truth than Rand, who knew everybody. For Sheldon was right about one thing: AIDS was rapidly turning into an inquisition in the gay community, as rife with terror and scapegoats as any launched by Rome. Does this one have it? Really? I heard it was his roommate. It gets so you can't be sick at all anymore. You hide colds, put Band-Aids on innocent bruises. You say you are fine when you want to scream.

I also fought for Roger's sister Jaimee, because I knew in my bones she'd be the rock my brother was proving to be. I'd only known her in the context of the family. She had a laugh as quick as her brother's, and his eye for the clay feet of power. I don't think Jaimee and I had ever spoken about anything major, but I knew she had an allergy to phoniness, as plainspoken and black-and-white as her father. She and Roger were effortless together. Jaimee could handle it, I told him.

Two years later she remembers that he presented it in just these words: *challenge, not a sentence*. No lie there. I feel a curious puff of pride to recall how gallant was our unity of purpose. Or perhaps I'm just glad that he took what I said to heart. When I heard myself say it, I didn't quite believe it. He was the one who forged a purpose out of it.

Fortuitously Alfred went off to New York for ten days, because it wouldn't have been easy to keep up the lie with him. We met nearly every afternoon for an hour of pitching. Richard Ide had already been given the thumbs-down sign by Marjorie Perloff, across a lecture hall at USC, but I only knew that after it was all over. It's typical of Richard's sense of honor—his leeriness of gossip—that he kept leaving messages in the two days following the diagnosis, asking me to call. This made me plead his case to Roger all the more eloquently, and thus we brought him in. So that is what I do now: I leave messages on machines, general and suitably vague, so as not to corner anyone. *Call when you have a chance*, I say.

Has anything ever been quite like this? Bad enough to be stricken in the middle of life, but then to fear your best and dearest will suffer exactly the same. Cancer and the heart don't sicken a man two ways like that. And it turns out all the certainties of health insurance and the job that waits are just a social contract, flimsy as the disappearing ink it's written in. Has anything else so tested the medical system and blown all its weakest links? I have oceans of unresolved rage at those who ran from us, but I also see that plague and panic are inseparable. And nothing compares. That is something very important to understand about those on the moon of AIDS. Anything offered in comparison is a mockery to us. If hunger compares, or Hamburger Hill or the carnal dying of Calcutta, that is for us to say.

My recollection of the two weeks Roger spent in 1028 is as fragmented as the weeks before the diagnosis. I'm not sure what preceded what, though I do have a sense of first week versus second, because the crisis took a turn midway that drove the panic off the graph. Roger was extraordinary in the days that followed the verdict, by which I mean he was utterly himself. There are fifteen separate client items in his calendar just for the two days after. I recall him bragging of that to Gottlieb with a laugh, saying that at five hundred a day for the room alone, he had to pay the bills somehow.

This force of life continuing is what they mean by "positive denial." When I first saw it in Cesar it bewildered me and made me fear for the crash landing. Now, as I watched Roger pick up and go forward, briefcase open on the bed, I felt how real and noble was the act of overcoming. I wanted to nourish that force in him and clear him room to horde it—hoping, too, that some would rub off. I quickly grew to hate being away from him, even for a couple of hours. It wasn't just the pretending that business was usual, though that alone was vile enough. In those first few days I seemed to draw my only strength from Roger. If he could do it, then I could.

Ernest Becker speaks in *The Denial of Death* of the heroism of doing anything at all in light of the mortal dilemma. I didn't start heroic, but it turned out there was no place else to go. I had coffee with Sam late one night, when he tried to explain my heart-pounding terror. I was flailing at the knowledge that Roger might really die, and yet part of me was still waiting to wake up so we could leave for Egypt. I pulled back rigid, defiant inside, wanting to drag Sam up there to see Roger laughing and working. This man wouldn't die; I wouldn't let him.

Besides—and here was the hero's corollary, drawn like a line in the dirt—if Roger died then I died. As my life blew into smithereens that week, a fund-raising letter arrived from Yale, and I wanted to scrawl across it: *Paul Monette died on March 12, 1985*. But Roger Horwitz didn't. I know because I was there.

Still, by the third day of treatment he was feeling dreadful, coughing and feverish. The doctors said it was typical of PCP to get worse before it got better, which sounded as mean as an old wives' tale. In any case, he had to channel all his energy to fight the egregious symptoms, plus the drug made him nauseous. Now I wanted to be there all the time—fuck normalcy—to handle the dealings with doctors and staff, to man the phone. It was just as well, since I became totally unhinged when I left the hospital.

I went to my own doctor to get a prescription for sleeping pills, and I poured out the unspeakable news. "What do I do?" I said in desperation.

He shrugged his shoulders with a cavalier unconcern I can only attribute to his certainty that he was safe himself. I've seen that straight man's shrug a hundred times. "Burn the blankets," he replied facetiously, scribbling a prescription for Halcion.

I felt as if I'd just been run over by a truck, while he went on to give me the benefit of his own pain. His brother had died in a car crash while he was still in med school. "Sensitive as a toilet seat," as Holden Caulfield says. I guess it was there in that encounter that I came to revile the comparisons of others. Is this how a Jew feels when he hears "holocaust" appropriated to some other calamity? Yet I was still so wounded by the news itself, desperate for allies, that I didn't have the wit to slam out of his office. Besides, I needed him just as Roger needed Cope, didn't I? I told him in some defiance that Roger hadn't exhibited the two *pre* symptoms the way he'd said, for two months running. I never stopped feeling betrayed by all those phantom barriers that hadn't worked.

"You live alone, you die alone," my doctor said sententiously, serene as Pilate. A month later I would overhear him fretting about what color Ferrari he should get.

*Not us*, I thought, as the rage began to build like a boiling tsunami. My determination to be with Rog every minute I could, whatever happened, took stubborn form in the doctor's office. White-hot rage is the only thing that keeps you going sometimes.

It was sometime that week I had a nightmare that dragged me awake screaming and in a sweat: I walked into the bathroom and looked around the shower door, to see Roger sitting in the tub, the water to his waist. He was dead. Head lolled slightly to the side, he looked like a dreamer himself, fit and healthy, his skin beaded with water as if he'd just come in from swimming. It was only half a second before I was roaring in pain. I reached in and lifted him out of the water. His heaviness, actual as the weight of a man—my own dead self—pitched the scream to a howl like Lear's as I lurched out to the bedroom, full of nothing forever. I'd never had a dream so physical or so desolating. Sam convinced me not to tell it to Roger, but I never came to terms with it. I'm still afraid it will come again, and have to remind myself in the middle of naps that Roger is already gone—I never sleep below the surface anymore—so the bathtub dream won't engulf me.

So many monsters have haunted the darkness of AIDS. Only four thousand had died by March '85, but already we all knew stories of men left incoherent in their own excrement, abandoned overnight by friends, shipped back to a fundamentalist family to pay the wages of sin. They were chained to their beds with dementia in New York. They lost their houses and all their insurance. The most horrible death in modern medicine, people said. This was the gossip for years, whispered among us with the same appalled prurience as used to be generated by the sexual exploits of the seventies. The latter of course still went on full tilt among straight people, herpes or no, because the full story of AIDS wasn't being told. It had stopped being somebody else for us, but not for them.

Gottlieb told us people broke through with either KS or an opportunistic infection, rarely both. The OI situation was much the bleaker. Wherever I went now, someone would come up and ask how Roger was. *Fine*, I'd say, *fine*. I had to negotiate several calls with a self-obsessive business friend who went bananas every time he had the runs or cleared his throat. One day I was crossing Santa Monica Boulevard and met Bruce Weintraub, who'd worked as set designer on *Scarface*. Bruce was thirty-three, an exuberant, intoxicating man who just before Christmas had moved into a thirties house in the next canyon over from us. I stopped to chat with Bruce, floated the good news about Roger, and we ridiculed the neurotic friend who was certain every anomaly was AIDS. I remember thinking as I walked away that Bruce looked washed out. He'd been diagnosed with KS a couple of weeks before, and was prepared to live with it as a secret, till he nosedived a month and a half later with PCP. KS and an opportunistic infection at the same time constitute the bleakest AIDS situation of all, and the shortest fuse.

I managed to be as positive around Roger as he was with himself.

In the afternoons we'd go over legal matters and household bills; then I'd drop a packet of work by his office in Century City, telling everyone there he'd be back directly, flashing a tight smile. I see now the only place I could fall apart was my work. Within a day of the diagnosis I became convinced it was over for good. My novel *Small Powers*, benign and didactic, my shelter from the earlier storm of Cesar, had overnight become an eccentric joke. I would not be going past page 223 after all.

This sounds ridiculously self-important, I know. I can only say it felt like a proper reaction to the life of self-importance writing itself suddenly looked to be. *Just the Summers* would never be put on now: no time and nothing to laugh at. All I could do was steel myself to finish *The Manicurist* so I could pay the bills, but the yucks were delivered with a skull's grin, and the whole enterprise filled me with loathing. Roger tried to help me through this minefield of self-denial, but I only dug in deeper.

Somehow it wasn't an option to write about the fire storm itself. For a while my journal stood frozen around the two-word entry for March 12: *the verdict*. No reason to keep a record of what was over now. Again Roger gently tried to steer me back on track. I should log my old journals onto the computer, something I'd wanted to do, just to be doing something. So one day I flipped on the PC and opened a creaky ledger to August '71. I read the first dewy-eyed page and wanted to throw up. Scratch that idea.

Of course I couldn't breathe a word to the producers and executives involved in *The Manicurist*. They weren't paying studio wages for hysteria, not in the urban comedy division. It happened that later that week I had a letter from the Alley Theater in Houston, praising a melodrama I'd sent to them months before with regards from John Allison. They weren't sure exactly how they wanted to proceed—staged reading, workshop production, main stage. I forget all the ramifications, but it was the sort of letter I used to megaphone to Rog the instant I tore it open, the rare Irish Sweeps among the pink slips. Roger was pleased as ever to hear about it, but I wouldn't move an inch. The play needed cutting, I'd have to be in Houston, and I couldn't go anywhere now. Sure you can, he said. Oh no, I couldn't. That part was finished.

If this all sounds like an excruciatingly unsubtle way of getting mad at Roger for being sick, I'll concede the point for what it's worth. My anger was surely growing more and more unmanageable. But I thought I understood the difference—then, anyway—between being mad at him and being mad at AIDS. Sam was persistent, saying that Roger must have a towering rage in him about being ill, and my own was just as understandable. Moreover, the anger was useful because it drained off a lot of the stress and would help me voice my most demented thoughts and thus get them behind me. Sam wasn't waiting for me to discover the answers here. We had reached the triage stage of overload, and he helped me work my way through the whole scenario. Roger and I must both get over the horror that we might die so we could begin to see we weren't dead yet.

Easier said. The only one I felt like being angry at was Paul. Self-recrimination filled up each successive vacuum as I tossed out work, friends, books, the world at large, one thing after another. The through line of my guilt, as an overdetermined actor would have it, went back to Joel and the unhappy months of 1981. If I hadn't had the deadborn affair with Joel I wouldn't have collapsed the way I did. If I hadn't been so full of havoc, Roger would never have gone east alone in October, never gone home with the freshman lawyer he met in Cahoots after seeing *Nicholas Nickleby*. He'd come back from that trip to the run of ambiguous viral misery that ended up misdiagnosed as amoebas.

We both agreed it must have been that contact. Since Cesar's diagnosis, I'd brought up the idea three or four times that as long as we weren't exposed during the bad patch in '81 we were home free. I think I said this not with dread but rather to prove what a needle in a haystack AIDS had to be to get us. Magic, magic. No way of knowing with certainty if any one encounter was the source, though it's remarkable how many I talk to have a sense of who and when. In any case, Roger expended none of his spirit on guilt or shame, ever that I could see. But I made up for both of us in spades, this despite his saying again and again that I mustn't look at life that way. "Paul, we got through all that," he would say about Joel, quieting me down even from his sickbed. Fate was the issue, if anything; not guilt. But at the beginning I wanted to die of the squandered life that was all my fault. What if I hadn't met Joel? Would Roger be well now? It took me months to work through the scalded cosmology of who and when. It took two of us, the sick one soothing the well one.

Yet that is a false picture of what life on the moon was like, hour by hour in room 1028. Though Roger slept a fair amount and curled up to wait the infection out, a lot of plain talk passed between us as we strategized this alien place. The ordering of meals, the hierarchy of staff, the byzantine complexity of the numbers. It all required a sort of continuous Platonic dialogue on the nature of us and them. Then I'd massage his neck or his flat feet—he used to say he fell in love with my arches—and we'd be as close as we were at home late at night, when everything was shorthand.

A hospital room is above all else intimate; you have no choice. Bullshit stops at the door, and the Hallmark sentiment of gratitude for the present turns out to be right on the mark. A precious ordinariness is the province of those who have loved a considerable time. So long as you are together, foxhole or detention camp, you make yourself a corner and make it work. In this enterprise we were fortunate to have privileges: we knew the right people and had enough money. Consequently we had a rare experience throughout of the very best of medicine, but especially here at the level of primary care, nurses and physical plant and patient skills. Ten East wasn't swamped and never had the uncontrollable feel of a state of emergency.

My friends in New York talk about food trays left in the hall because no one will bring them in, nurses with shaking hands, masked and gowned like astronauts, sweat beaded on their foreheads. We, too, were under the weird strictures of isolation, requiring ornate procedures for bagging trash and taking specimens. But almost nobody acted spooked at UCLA. I watched one or two nurses wrestle with it and get to the other side. As a fierce IV nurse put it eight months later, recounting a lecture she'd given a conference of her peers: "If you don't like AIDS get out of medicine, because this is where it is." That was the front-line sentiment in white, and otherwise we had the incalculable safety of a private room. The privacy alone gives you light-years of time.

Then, just as we were getting into a sort of rhythm, when it seemed we'd weathered the shock and the brute intrusion of medicine men, everything took a turn. Roger had been on Bactrim eight or nine days when he started to spike a fever, 102 and no explanation. They may have mentioned it could be a drug reaction, but the interns tended not to hazard a guess out loud. They scurried away and paged Cope. The fever put Roger flat out, and I went into a tailspin, sure the drug had stopped working and something new had flared. I know I risk the tale growing hoarse whenever I speak of a new level of terror and anxiety. But if I had an objective measure, on the order of the Richter scale, that fevered day in March was an 8.0.

While we waited for several consulting physicians to have a look, I must've been doing some paperwork to keep from going nuts. Roger and I had seven separate bank accounts, and somehow we fell to talking about me not knowing what half of them were. He began to explain how they worked—methodical, patient—and I scribbled notes on the check register till I realized this was what I would be left with if he died, just notes on being alone. I started to choke with tears, and Roger said from the woozy peak of his fever, "I guess you don't want to do this right now."

Kathy Hendrix had asked me to dinner that night, and since she lived a scant ten blocks from the hospital I ran over there while Roger dozed. I had to make contact with somebody I could trust, and I couldn't wait till midnight, when I began the long insomniac hours talking to Sam and my brother. Kathy is a reporter for the L.A. *Times*, and if there's a common thread to her work, it's the issue of conscience: no priest crosses a nuclear line in Nevada that Kathy doesn't cover the event. I walked in stark and staring and talked nonstop for an hour, spilling the secret and everything else I could think of, as if to break a fever of my own. I remember I asked her if she prayed, and she, who is no more religious than I, tried to compose me an answer. What was good about prayer, she said, was how it centered the mind.

I careened back to the hospital and found Roger asleep in the semidarkness. The enormity of our loneliness gripped me like angina as I looked out the wide plate windows over the city, west to where the lights froze on the ocean's rim. I stroked my friend and talked gently to him. I don't know if that sense of vigil went on for two days or three, but the helplessness was unendurable. And when early one morning Roger called to say a man from Infectious Diseases had been in and detected a faint rash on Roger's chest, pointing to a drug reaction and nothing more, the wild relief surged between us like adrenaline. We laughed as if we had heard an old joke from a distant country. The full roller coaster overnight—we experienced it eight or ten different times in the course of the war. Each time it was as if he came back from the dead.

They took him off Bactrim immediately. We lucked out here, because a new drug had just come on line to treat pneumocystis. Pentamidine was so new it wasn't listed in any of the reference systems yet. Even the hospital pharmacy couldn't find it on the computer. As soon as Roger went on it the fever disappeared, and he began to get better again. I didn't think then of the thousands in the first five years who turned out to be allergic to Bactrim before there was something else. I guess they all just died. They didn't get the nineteen months' reprieve that Roger and I would hoard coin by coin.

An offensive strategy began to emerge on the island of 1028, especially as I took an increasingly hands-on role, pestering all the doctors. Together Roger and I became postgraduate students of the condition. No explanation was too technical for me to follow, even if it took a string of phone calls to every connection I had. In school I'd never scored higher than a C in any science, falling headlong into literature, but now that I was locked in the lab I became as obsessed with A's as a premed student. Day by day the hard knowledge and raw data evolved into a language of discourse.

We were Cope's pupils first, of course, and he understood immediately that we wanted the closest involvement in the process. Every intern, every nurse, was detained for a lesson: on catheters, lymphocytes, antivirals. Eventually the interns had more to learn from us than we from them, for we had a data base larger than theirs. I had taught school for the ten years of my twenties, and I knew the gold stars always went to the most skilled balance of brains and charm. So I charmed the pants off everyone. I might not ever sell another story, but I sang and danced up and down 10 East like Zorba at a wedding. Nobody left that room without being engaged by one or the other of us. Picasso once said that if they put him in jail and took away all his paints, he would draw with his spit on the prison walls. At our best, Roger and I were the talking equivalent of that irrepressible force. Tell us, we said, tell us everything there is to know, and we'll do the rest.

Sometime during the second week Jaimee and her husband, Michael, arrived in Palm Springs. On Saturday they left the kids with the grandparents and flew up to visit Roger. Once again we agreed I would stay away, since Michael had not been told the truth. Was I such an open book as not to be trusted to carry it off? Didn't I field half a dozen calls a day and calm the multitudes? I was the Gielgud of reassurance. So it wasn't that Michael would divine the truth from the hunted look in my eyes, but rather that Roger chose to take the field himself and give me a breather. All alone on that hot Saturday, the smog like a film of milk on the city, I kept calling 1028 from pay phones as I ran my aimless errands. When I stopped to get dog food at Hughes, I started to cry in line. Supermarkets are bad for grief; any widow will tell you.

Jaimee phoned from Palm Springs nearly every day thereafter, relating how she would squirm at the pool with her parents, listening as they talked about Roger regaining his strength, relentless on the bright side. That's what drew Jaimee and me so close, the smile we had to flash to the world at large. With one another we could let it crack.

We had already targeted Tuesday for Roger to come home, since Cope and the other doctors felt the third week of treatment could be administered by injection rather than IV. A visiting nurse would have to come in once a day. Much gearing up was required to engineer the change of place, but Roger was chafing now to get out of there, and the eagerness alone appeared to quicken his recovery. He was pale and exhausted from the battle, with an angry sore at the corner of his lip that was diagnosed as herpes and treated locally. But I found myself responding in kind to his air of urgency, feeling a rush of renewal at the thought of bringing him back to full recovery. I ached to fatten him up.

Yet this sense of being equal to the fight went hand in hand with the blackest conversations every night—my brother at midnight, Sam at 1 A.M. The actual suicide thought came during the vigil of the Bactrim fever. I was home from the hospital every day between afternoon and evening to feed the dog and rest, clenched and wide-eyed on the unmade bed. I would cry for a while with bitter despair, till one day I was struck with the notion that I could make the pain go away by dying.

No, wait: I had to be here for Rog. It became unthinkable the moment I thought it. Yet at the back of my mind the easiness of death stayed with me, mine if not his, in the raging blanks of the next two months. When you are gay and alone and want to be a poet, suicide crosses your mind at twenty-two like an impresario's cape. It is not the real thing, any more than the lightly stroked razor Joel had drawn across his wrist at New Year's. By contrast, this particular kiss of death was as lucid and merciful as putting down an animal in pain.

Over the weekend before Rog came home I called both Cesar and Craig and told them what was going on. I realize I don't sound very good at secrets, but I'd neglected to call either of them in over two weeks—unheard of. I hadn't even responded to Craig's letter announcing his diagnosis. Because both were being highly selective themselves as to whom they told, because both had a certain awe of Roger, I knew they'd respect his silence. I also needed to let them know I couldn't be counted on for optimism anymore, or to be a main line of support as I had been. My arms were full. And another reason, still only partially formed: I could lie to everyone out there, but not to my fellow exiles on the moon. Within three months this sense of separateness would grow so acute that I really didn't want to talk to anyone anymore who wasn't touched by AIDS, body or soul.

Cesar was on his way to Uruguay, to visit his family over spring break. It was his turn to do the bracing and buoying up as I fell apart, and he spoke with a conviction that wasn't feigned. "Paul, he's going to get better, you'll see. Don't be afraid." You only listen to such bravado from those who are there. Other people say it, and you stare out the window and wait to get away. You prefer the despair to hollow shows of strength. "We're all tough," Cesar said fervently. "It's not so easy to kill us."

Craig, on the other hand, was my research associate. He had two years' worth of the epidemic behind him in New York. There the death toll was running amok, like Flanders in 1916, while we in L.A. were still flinching from the gunshot at Sarajevo. What was the longest anyone had survived after PCP? What were these drugs the gay press was making noise about? Craig was my conduit to the AIDS underground, where every rumor was run to earth and codified before the mainstream press ever raised its head from the mire.

Monday night before Roger's release was Oscar night. I ordered dinner in 1028—another of the decencies of 10 East: $14.50 for a guest meal, and they took Visa. I got there in time to see Dennis Cope, who said everything was set for tomorrow, and meanwhile who did I think would win? Win what? I had to backpedal to recall what civilians were doing that night. I went into an automatic routine about the Nobel level of self-love unleashed by the Oscars, which made Cope laugh. Then I handicapped the race for him, earnest as Siskel and Ebert. We were actually talking about something else for once.

Roger and I had a lovely lazy evening watching the show, hooting at all the clunky grandeur. On the way home after midnight, exhilarated and scared about how we would manage by ourselves, I was stopped by a squad of cops on the Strip. They were holding traffic at bay while a line of six limousines turned in at Le Mondrian, Prince and his entourage in the lead in a purple stretch. I watched numbly as the glittering prom queens poured from the limos. Behind the tinsel in Hollywood, they say, is the real tinsel. But who were the moonfolk here—we or they? I felt it as a kind of physical pain, to think that life on the surface still went on in its gaudy rounds.

Next day Roger and I bristled with anticipation, as we waited impatiently for the pharmacy to send up the medication to see us home. For two hours we cooled our heels, until a young woman walked in carrying a shopping bag of drugs. Explaining how unfamiliar the pharmacy was with Pentamidine, she showed us how to have the visiting nurse mix it. I don't think I registered that the nurses wouldn't have even heard of it. Then there was an unguent for the herpes, a tablet to suck for the spot of thrush on his tongue, and a supplement of some amino acid because an AIDS patient's body no longer manufactured it.

When she left we stared at each other, and Roger gave out with a groan, suddenly overwhelmed. We weren't going home scot-free at all. But you either turn a moment like this into a black joke you can laugh at, or you'd never get out of 1028 at all. When we were told we would now have to wait for a wheelchair escort to take us down to the parking garage, we crowed our independence. Roger declared he was strong enough to go on his own power, thank you. We gathered up the bags and went down ourselves, laden the way we used to be in airports. As we left the elevator and made our way through the crowded lobby, I had my first experience of protective helplessness, lest someone should sneeze in our direction. I wouldn't let Roger touch the door going out.

As we drove home along Sunset, he sighed with pleasure at the open window, feeling the green of spring rush by. There are moments of reprieve that are happier than anything else in life. It's true the homecomings tend to merge as well—the dog turning himself inside out squealing, Roger making his slow way up the steps past the coral tree in front, collapsing with exhaustion in the bedroom by the pool. That's the best room to convalesce in, with the morning light and the bougainvillea. Our own room was shaded by the coral, perfect for my late rising but too dark for somebody stuck in bed. What I remember most clearly about that trip home was Roger sleeping the first twenty-four hours, practically straight through, so that all my plans to stuff him with baked potato and protein shakes went into abeyance. It was partly sheer exhaustion from the transition, partly the codeine, which he had to take to ease the pain of the Pentamidine injections.

For they'd made a mistake sending us home with intramuscular instruction. Pentamidine doesn't diffuse into the veins very well. It goes in like a protracted hornet's sting and collects in a painful bubble between the muscles. The discomfort was so great he needed his Tylenol laced, which nodded him out the whole next day. I'd wake him up every couple of hours to make sure he was all right, frantic at realizing we were all alone, with no hospital staff to ask for advice.

But in fact it was an extraordinarily peaceful sleep. Roger kept waking with a smile and telling me not to worry. It was such a luxury, he said, to doze away the day in his own bed, uninterrupted by the hammering routine of 1028.

But within two days we were begging Cope to let him come off the Pentamidine. By then Roger had welts from the shots on both cheeks of his buttocks, so sore that he had to lie on his side. The last straw was the Saturday dose, administered by a rattled male nurse who injected 2.5 cc instead of 2.0. Roger gritted his teeth against the pain as I ran out to Thrifty to buy a heating pad. It was not the last time I wanted to open up with an Uzi in the long line at a drugstore.

Yet when I wrote that night in my journal, the first dispatch from the moon, we had managed to leave the mess of the treatment behind us. Apparently even I could sometimes see the bright side:

> *Lying next to Rog in the guest bedroom. We went out for supper to Cock and Bull and took a walk down Cory St. I never thought I'd write in here again, I never thought I'd do anything again, but I record with gratitude and a sense of calm that we stepped out tonight for a plate of prime rib.*

Here it breaks off because the jiggling of the pen was bothering him. In fact we were both rather purring, maybe even a trifle cocky, having the evening off like that. It is something you never expect to be a great strength, the talent for small pleasures. Of course it only gives you the inch; the mile is another matter.

Cope finally allowed us to stop the drug, on the fifth day home. By now Roger was out of bed longer and longer, champing the bit to get back to work. Though I thought it was too soon—he was still so gaunt and weak—we made plans for him to start going in at least part of the afternoon. Meanwhile, on April 1, Al and Bernice finished their two months' sublet in Palm Springs. As they planned to be in town overnight, we invited them for dinner, which Bernice insisted on cooking. Roger was lying down when they came, and his mother went right to the kitchen to put together a meat loaf. I remember getting Roger up, and as we came through the study to the kitchen door he was right behind me. Bernice stood at the sink with her back to us, kneading the ground meat, when suddenly I realized Roger had disappeared. I followed him back to the bedroom, closed the door and found him in the bathroom, weeping.

"My poor parents," he cried as I cradled his head in my arms. I rocked him and soothed him and said the right thing, but now I only hope I let him cry enough. Considering the sea of tears that I produced, Roger never cried much at all. Even then it was only a wail and five seconds' squall, which always made the occasion that much more intolerable and wrenching.

Once he'd composed himself, it turned out to be an easy family evening, plain as the meat loaf, and next day the parents left for Chicago, satisfied we were on the mend. I began driving Roger over to Century City in the afternoons. For the first few days I hovered there, as if I might be needed to keep people from getting too close. I was like the hall monitor, trying to root out who might have a cold among that whole suite of attorneys. Roger had been doing work for a couple of high-powered types in the suite, both straight. AIDS to them was still page 48 of the second section, so I don't think they even wondered. They liked Roger and were glad to see him back, and meanwhile when could he have the documents done?

Within a week I was leaving him off outside the Century Park towers at one and picking him up at four-thirty or five, with several frantic calls between to make sure he wasn't too tired. He was so glad to return to a semblance of normalcy, though it frustrated him not to have more energy. He had two lawyering friends, Ackerman and Comden, with whom he was glad to be joking again, bemoaning the plight of the sole practitioner. I remember that time as so peaceful now, waiting in the Jag at five o'clock, parked at the curb like a Connecticut wife, as various three-piece men and women revolved in and out of the building, hyper with energy, meters ticking. At last

Roger would emerge, looking a little rocky, but with his tie neatly knotted as ever and his briefcase firmly in hand. I felt such a flood of love for him then, and wagged like the dog and happily chattered as I drove us home to safety.

I have the evidence in hand that I laid down two or three pages of *The Manicurist* each day, but I scarcely noticed. In fact I would sit at the desk wanting to leap out of my skin, counting the hours till I could go pick Roger up. The role of the solitary scribe had become insupportable, but Alfred and I had nothing in the works together. And they were breathing down my neck at the studio. One of the rabid executives—they of the twenty-hour barracuda days—scheduled a 7 A.M. conference call, where I was shrilled at and all my pages spat on. I just watched the clock and waited to go back to bed next to Roger. I realize it isn't a matter of great suspense whether or not *The Manicurist* ever got made. It was clearly doomed from word one. But at the time I was under the same gun as any writer with a high concept: *Get it done now, make it like everything else, and this time make it funny.*

My real work, as far as I was concerned, was to find out all I could about this disease. I talked to Craig nearly every day and thus kept up with the early word on HPA-23, the antiviral drug at the Pasteur Institute that was starting to draw to Paris desperate men from the U.S. If the main thing that got Roger on his feet was getting back to work, what fired me was tracking down a cure. It became a kind of compulsion, and gave me the best shot at a positive attitude as I reported my findings every night at dinner. Besides, it was all we had to hold up against the fearful statistics. We had a follow-up appointment with Gottlieb in the AIDS clinic at UCLA, during which Roger asked him how well his other patients were doing. Gottlieb said with a certain pride that he'd kept some alive as long as three years. The stricken look on both our faces was exactly the same. Three was the *good* news?

It was a constant battle against doom and gloom, as each new journalist stumbled morbidly into the ravaged arena. In April an article in *Rolling Stone* quoted the records of a monastery during a plague in the Middle Ages. The single entry for one year—in Latin, one assumes—was "More dead." I didn't read the piece, but Roger perused the first page of it, waiting for me to get ready to go for a walk. As I came into the study he let the magazine fall to the floor and said in a quavering voice: "I don't want to die."

"You won't die," I said forcefully. "You can't die." Always there to buck him up, if not myself. We'd follow every lead, I told him, and be knocking on the right door the minute they found the answer.

We always took our walk up Harold Way, which starts across from the house and runs along the brow of the hill from Kings Canyon to Queens. Both are steep box canyons, sparsely built because of the precipitous angle of repose, and covered with a tangle of chaparral as old as the mountain it grizzles. At the bend where Kings turns into Queens Canyon, out on that point is where Liberace used to live; he'd brought a brief flurry of publicity to the neighborhood when he tried to convert the house to a museum. That was before our time, and he'd settled on Vegas for archival purposes, but the wrought-iron gates still bore two curlicued L's. Those were the days when the Hollywood Hills were known as the Swish Alps.

We must have taken that walk a thousand times in the six years we'd lived up there, but especially late in the evening. Often we made it a full circle by hiking uphill to the next street parallel to Harold, where there was an empty lot covered in century cactus, with the full *Star Wars* view, from the San Bernardino Mountains in the east to Catalina in the west, maybe seventy or eighty miles wide-screen. We'd had so many temperate moments there as we lingered to drink in the view from the eagle's perch. The first couple of months after the verdict we could still go the full walk every now and then, but at least some portion of it was his major form of exercise as Roger worked to build his stamina again.

One day early on, we were pacing ourselves up Harold Way, and the city below was smoky blue under the marine layer, the so-called Catalina eddy that is common on late-spring afternoons. Long before there was smog, the Indians had called the place Valley of Smokes. "Isn't it beautiful?" I said with a rhetorical nod. I think what I really meant was how delicious it felt to be walking together again. Roger stopped, still a bit hunched with fragility, and looked out over the city: "But this is the place where I got sick."

And I was stabbed with a certainty that none of it would ever have happened if we'd never left Boston. But enough of my guilt for now. I mention the moment here only to show he was as capable as I of such stark and devastated observations, in which the cup is bone dry. Yet I note these two especially—*Rolling Stone* and the April walk—because they were so rare. As it took a great deal to gather Roger to a burst of tears, so, too, he gave vent to desolation in the single throb of existential pain. A few words were all he needed to speak it clear, no matter how much we might say afterwards to make it go away.

Sometimes weeks would pass between such melancholy briefs. Did he therefore keep it all in? Of course I can't say for sure, but here his Athenian balance served him well, in that he took it a concrete step at a time. An hour of work, a concert on PBS, a visit to friends we could laugh with—friends who hadn't a clue. He could lose himself in things in a way I never could, not till much later. But if I have any sense at all of how we persevered so long, it comes down to an equal measure: an unwavering goal to beat it, and the group of two for an army. In combat Roger had no choice but to battle the physical side, while I engaged on the metaphysical front. A simplistic formulation if you take it too far, I know, but it took us further than either of us could ever have gone alone. "The pals," as Roger used to call us, nudging me shoulder to shoulder.

Sometimes we'd go to the Detroit Street building together, and I'd be maddened by the Sisyphean tasks of the place, while Roger puttered about and got them done. Right after he came home he was fiddling with the plumbing one day in an empty apartment, and crouched by the toilet to get a closer look. I freaked out. The place was grubby and crawling with microbes. I wanted to wrap myself around him like a bubble, my need to protect was so desperate. Already I'd started to keep the surfaces of life insanely clean, wiping the phone and the doorknobs late at night with a cloth steeped in ammonia.

Yet there was an evening when we flared into a fight because I couldn't remember if I'd run the dishwasher or not. I said yes, then Roger took a glass out, then I said no. And the tension broke with our nerves all jarred. There was so much panic just beneath the surface, in a world where a single unwashed glass could kill. Thus there was a level of protectiveness that really didn't want him to go to work at all. What were we doing going back to normal? The only thing I wanted to do was be with Rog. I seemed to have no other plans.

Then he began to get better for real, the stoop went out of his walk, and I had no choice but to let go. There was a great light-headed moment when he was washing up one morning and suddenly called me into the bathroom. The herpes scab at the corner of his lip, so black and crusted we had to put a Band-Aid on it before he went to work, had finally shed itself. It had been a last lingering reminder of the world of 1028, and we rejoiced to see it vanish. That is how minute the sharing is, how private the victory—someone to show that your scab is gone.

And if Roger was getting better, then I simply had to do something about the cloud of death that shadowed me. My appointment with Sam on April 4 was in the nature of a lecture on how to get over an opportunistic infection of the spirit. I'd clearly not accepted just how powerless I was, he said, and was stupefied with rage that I couldn't command the internal workings of our two bodies. Nevertheless I must stop qualifying life "if Roger gets better" and start asserting *when*. And I must refuse to let us go quietly. Fighting was to despair what aspirin was to fever. Stop living in a state of premourning, Sam said.

I must already have pulled together some information about going to France for treatment, because I spoke of the awkwardness of providing cover. Neither set of parents knew, so how were we to explain going over to Paris for the summer, when Roger was supposed to be back at work? Sam replied succinctly: Who cared what they thought? No stone must be left unturned, no matter if it took us to Tibet. And even as I made whatever radical plans were necessary, he advised, I must also gear up to get back to some mindless routine. Have people in again. In short we must restore ourselves to our life, whose character had always been the opposite of morose and doomed.

We were about to join a community of the stricken who would not lie down and die. All together, we beat down the doors of the system and made it take our count. Some have sat in medical libraries wading through the arcana of immunology. Others pass back and forth over the border, bringing vanloads of drugs the law hasn't got around to yet. This network has the feel of an underground railway. It could be argued that we're out there mainly for ourselves, of course, and the ones we cannot live without. But on the way we have also become traders and explorers, passing the word till hope is kindled in places so dark you can't see your hand in front of your eyes. If the government was going to continue to act as if we didn't exist, if the medical establishment was prone to gridlock over funds, if the drug companies were waiting till the curve got high enough for profit, then we would find our own way. Whistling in the dark is whistling still.

We had been to the brink in March. Now that Roger was home, we had a window to let in air and a certain breathing room to fight. No time to waste, because no way to gauge how soon the window might slam shut on our fingers. My own hands flinched and balled into fists when I typed that line, recalling an afternoon in the spring of '75. Roger had just moved into the apartment on Chestnut Street in Boston, and was cleaning windows in the living room while I shelved books in the bedroom. Suddenly he shouted in pain, and I ran in to find him trapped, his two thumbs jammed by the heavy window because the cord had severed. I still remember the sickening guillotine feel of the sash as it came away from the flesh, and engulfing him in my arms as his thumbnails flushed dark purple.

Yes, we'd decided to fight. No, the despair wasn't gone. The two emotions jockeyed in our hearts. You had to be there all the time to know which was dominant in a given hour, a given minute—the clock doesn't parse fine enough to tell how vast and swift the mood swings were. But if you have ever freed someone from pain, you know why it is that a mother can lift a car off her trapped and whimpering child. Give us then the bravado of days when we swore we would beat it, for underneath we were scared as ever, and always pleading silently, *Don't let it come again*.

# ·V·

*4/11 Wednesday*

*The closest I came to believing something higher—after the loss of the old Episcopal thing—happened in Greece, and centered on the Greek ideal: scholar, philosopher, athlete, warrior, citizen . . . it gave me a context. But how is that context still valid, when it seems like it only fits the joy of intensely living as R and I have been doing over the last years, all the Greek parts in flower. What's* left *of that ideal? Just Greek tragedy, the horrors of fate? How to be a Hero—the thing the Greeks believed in most.*

There wasn't all that much to know in April '85. The first drug anyone knew by name was HPA-23, and the first person in our orbit to go after it was Tom Kiwan, a lawyer who lived a couple of blocks above us in the canyon. We didn't know him well, but Alfred was his neighbor, and Tom was at a stage of panic that gripped people by the lapels. Along with hundreds of others, he'd been monitored for a couple of years by the gay men's health study at UCLA, his blood work updated every few months. In April a doctor told him his numbers were in the red zone. He also had thrush on his tongue, an ominous sign. There are doctors who now consider thrush evidence of full-blown status, but in the spring of '85 the sliding scale of definition was still drowning in backlog.

In any case, Tom wasn't waiting around. He flew directly to Paris to check out the operation at the Pasteur Institute. Fortunately, he and his lover spoke French, and were able to arrange for Tom to enroll as an outpatient in the HPA-23 study. The inpatients were living in barracks, many of them all alone and without a word of the language. For some it was literally the last ditch, a secular Lourdes. Tom would go back for the full two-week course of shots in May; he would be dead by Christmas. I can't assess what time the drug may have gained him, but his story went into the pipeline like a crude-drawn map, explorer division. I know what a boost it gave us all to hear that someone was charging ahead. You run in the steps of the hunter before you.

By midsummer the world would know that Rock Hudson had been treated twice with HPA-23, but by then the news wasn't any use to the pipeline, which was scrambling for information about the next generation of drugs. Because Roger and I had our own secret, I'm in no position to criticize anyone else's profile. But all along I made sure the circuitous route of our search for the magic bullet got out to the AIDS underground, even when I had to deep-throat my source and say it was some vague "friend" receiving the drug. Neither do I blame the rich and well-connected for chasing cures available to them through the hierarchy of Who You Know. After all, Roger and I would never have muscled our way into two experimental programs without our own friends in high places.

Yet it's still very difficult to accept that men of our tribe succeeded in obtaining these elixirs, however worthless, while the rest on the moon were clamoring. Long before Roger got sick there was a persistent AIDS rumor about one of the major players in the fast lane, a zillionaire I'd trailed through the club scene in New York while researching a script in '83. He supposedly went to Europe every other month to have his blood replaced. It's hard to separate such exotica from the monkey-gland youth search that has always landed the well-heeled in the clinics of Zurich. The seductive twist to this tale had it that the man regained his health completely after a year of very dicey symptoms. I know we can't all go to Switzerland; the mind reels at the cost of *that* private room. I'd just feel a whole lot better if I knew for sure exactly what it was I couldn't afford.

Roger struggled through one drug study for four months till it almost killed him, another that gave him nine months' grace. They were closed systems that asked for anonymity, that wouldn't have been able to expand their resources no matter how loud people banged on the door. We came into the research wing of the nightmare at the earliest stage, with the merest handful on the lifeboat. But once inside, we pleaded for everybody. Why was it taking so long? If they had it here in this room, how could they keep it from those who had no other hope? What business was it of the FDA's?

Our own power source was a man with impeccable connections across the board, from Sacramento to Washington. He also had status as an academic insider in the UC system, such that his calls were returned the same day by everyone from the president on down. He and Roger had known one another since Roger was a kid, and there was never any question but that his full clout was at our disposal. Behind the high fence of experimental drug research he was tireless and unfailing, a mover of mountains.

And he understood that Roger and I were not going to wait; if we had to go over to France to get started on something, we would. In hindsight I wonder if perhaps we were precipitous. What they call the "honeymoon" after the first infection—when a person with AIDS will often feel better than he has in months—could well have gone on for a year or more. Nothing said Roger had to be on an experimental therapy within eight weeks of his hospitalization or else. Indeed, now I see how innocent we were about just how uncertain experiments can be. Bred as we were to literature, perhaps we leaped too quickly to happy endings. But every clock was ticking like a bomb. Having watched so many die without a chance, we couldn't let it slide. Death was close as the wall of this room, so what did we have to lose? Besides, courage doesn't precede action. The action releases it, like endorphins in a marathon runner.

I'd think sometimes how radically altered Paris would look when we got there, filtered as everything was now through the gauze of mortality. Over and over I'd picture the bullet holes in the stone walls on the Île de la Cité, where Resistance fighters were gunned down in the street. Here and there, beside the scar in the stone where a bullet had struck and ricocheted, was a small marble plaque: *Ici tombe Jacques Vassal le 12 juillet 1944*. Here one man had fallen on a particular summer's day. If life was pocked like the walls of Paris now, at least it would be Paris, and Madeleine would be there. We wouldn't have to live in the barracks.

Then our power broker put us in touch with a researcher at UCLA who had access to an experiment about to start in Immunology—a different drug from HPA-23, as good and perhaps even better. I'd be on the phone to Craig at every juncture, feeding him data, because he had two high-placed friends as well, one a distinguished immunologist in Houston, the other a doctor in Stockholm. For a while I couldn't be sure if we were talking about one drug or four, but I was beginning to grasp the notion of what "antiviral" signified. All the great breakthroughs in the antibiotic department were marshaled against bacterial agents. There had never been a magic bullet for viruses, and AIDS had come along just as some small progress was being made. Or as my Ferrari doctor was wont to tell me, AIDS would prove in the end to be a great boon to medicine.

Mid-April was also the first international AIDS conference in Atlanta, the most public feature of which was Health and Human Services' Margaret Heckler, announcing that the Reagan government was going to do something at last, now that the disease was a threat to "the general population." The conference added precious little new to the body of evidence. It was split along French and American lines, the prima donnas still squabbling about who invented HIV. Priorities, please: there were royalties to be protected.

By the beginning of May we had the four drugs straight in our minds. Besides HPA-23 you had suramin, foscarnet and isoprinosine—I know these words now the way I know Alka-Seltzer and Bufferin. Suramin, which had been around for decades, was the breakthrough drug that first successfully treated sleeping sickness. There were extraordinary accounts, more vivid with every telling, of the Lazarus waking of thousands of slumbering victims when suramin was introduced in the twenties. Like suramin and HPA–23, foscarnet was an antiviral. It was being tested in a hospital in Stockholm, where Craig's doctor friend happened to be affiliated. They had four patients on it in Sweden—four in the whole world, that is—and it seemed even more of a gamble than Paris. They hadn't even set aside the barracks yet in Stockholm.

Though isoprinosine was being touted as an "immune booster," right from the start several doctors we talked with discounted it as junk. As one of them said about a later cure-of-the-day: "You might as well drink your shampoo." Nevertheless, isoprinosine was available over the border in Mexico, and already there was a pony system set up to go fetch it in Tijuana. Suramin sounded to be the most promising of the lot, but in truth the promise was based on almost no hard data at all. More to the point, suramin was the drug whose efficacy trials were set to go at UCLA.

I've here collapsed a process that took eight weeks between knowing nothing and Roger's getting his first dose, because the struggle for the drug gave us a great surge of purpose that colored everything else. Any news about any drug could cut through my blackest despair. Also, there was a sea change in our perspective as we heard the tale of the ragged band gathered outside the high walls, shouting and pelting stones. We came to understand just how deaf the collective Reagan ear had been in the first four years of the calamity.

And if the government was stone-deaf, the press was mute. The media are convinced in 1987 that they're doing a great job reporting the AIDS story, and there's no denying they've grasped the horror. But for four years they let the bureaucracies get away with passive genocide, dismissing a no-win problem perceived as affecting only an underclass or two. It was often remarked acidly in West Hollywood that if AIDS had struck boy scouts first rather than gay men, or St. Louis rather than Kinshasa, it would have been covered like nuclear war.

In September '83, Cesar was circa case two thousand. By March '85, Roger was number nine thousand, give or take. In addition, there had to be one or two hundred thousand others suffering symptoms of AIDS-Related Complex (ARC)—diagnosed or not, people who just felt awful and kept getting sick. So the lucky few dozen who had a shot at suramin were the first to receive the slightest therapeutic glimpse of daylight. To be sure, hundreds in California were self-medicating with isoprinosine, and there were the restless travelers like Tom Kiwan. KS was being bombarded with chemo: Cesar went on vinblastine starting in January, and had already started to lose his hair. Even the rage for macrobiotic had begun, though many felt it wasn't high-caloric enough for a disease that tended to waste people.

But for all that flurry, I can't say strongly enough what a quantum leap the antivirals were—the possibility at last of treating the underlying condition. And here was a free ticket. I think we both felt a quickening of pride to be pioneers, and of course we thought the drug would work. One night in the middle of the intelligence gathering and string pulling, Roger and I sat with chairs drawn up to the stereo speaker, huddling to listen to a static-ridden tape. A few days earlier Craig had flipped on his journalist's Sony while talking to his research friend from Houston. The man discussed each of the four drugs, and though he continually stressed that none was a certain answer, Roger and I could not mistake the suppressed intensity of his eagerness to see them tested. Hunched like a couple of Poles or Czechs listening to Voice of America, Roger and I exchanged a wondering smile.

But the roller coaster didn't go away, even if I finally had something positive to occupy my afternoons between spurts of humorless script dialogue. And we did begin to open our lives again to friends and evenings out. No event was simply itself anymore, of course. But I look back at some of those evenings now, and to a camera eye they're almost the same as before the war. People didn't have to know it was AIDS to be there for us, with all their blissful sameness. On Friday, April 12, an actress friend with serious medical problems of her own, which wobbled her muscles and made it hard for her to walk, called and announced she was bringing over dinner. "I only make one dinner," she said. "Curry. So that's what you're getting."

Roger was very tired from the week at work, but we had a lovely evening with this lady, who'd always been wise about how to bear the sting of critics. Once we saw her do a Beckett play, a monologue in a rocking chair, that was a tour de force of concentration. In the summer of '84 she'd performed a monologue of her own, *Conjure Woman*, sitting on the edge of a stool, gesturing like a dancer. She still managed to act though she could barely walk, and thus she passed the test of those whom I would listen to at all these days.

After she left, Roger and I were in the bedroom watching the news, when a friend called to say the designer Angelo Donghia had died that day of AIDS. I'd never met Donghia, but it struck me even then that here was a secret that didn't go public till he was gone. How closed off had he been at the end, I wondered. The shock of the sudden deaths is still disorienting, especially among the celebrated. It's like they're snatched out of a chorus line with a hook.

And the obituaries are telegraphed instantly through the underground. Part of this is prurience, of course, but more it's the need to say out loud what the press and the wary so often blur: *died of pneumonia . . . leaves a sister and a Persian cat, Meow.*

Because Roger was lying right beside me I couldn't not tell him about Donghia, though I quickly learned there was no need to drumroll the names of those we didn't know. We had enough of our own. Roger was always visibly pained to hear about anyone dying. He wondered now with bitter dryness if Donghia had made it the full three years.

Yet in spite of the late bulletin, I note in my journal that we ended the evening by floating the notion of being grateful for a good day. Before he turned out the light, Roger said with quiet amazement: "I'm really looking forward to tomorrow." As if it were any old Friday night, with all the bourgeois pleasures waiting on the morrow, from Koontz Hardware to the County Museum.

Saturday night Kathy Hendrix had a send-off party for her friend Camille, who was heading for a year in London to write a novel. It was to be our first occasion since the hospital bigger than two or three people. Somehow we had to balance our trepidation against the need not to build a prison around us. As it happened I had to leave Roger off at the door to Kathy's building in Westwood, because it's surrounded by fraternities, and the parking's insane. I ended up having to park the car five minutes away, and I ran back to Kathy's in a fevered panic, horrified that Roger was moving through that crowd defenseless. What if somebody kissed him?

I never quite recovered my equilibrium, especially when I saw there were thirty-five people milling about. Roger sat on the sofa and chatted with Jill Halverson, director of the Downtown Women's Center, whom some of us call Saint Jill, though she is far too wry and irreverent for beatification. Roger was doing the legal work for the resident hotel that Jill was renovating for the women of the street. Meanwhile I wished Camille good luck with her novel, trying to push away the thought that we wouldn't be here when she returned the following spring. She spoke of the flat she'd be living in near the

Thames, and it was as if I kept drawing a black curtain over every green and civilized English memory. Ten months before, Roger and I had wandered in St. James's Park on the blooming day before the Trooping of the Color. That was the end of England; nothing to speak of there anymore.

I bumped around the party, hovering over Rog, not knowing how to talk to people who knew me before the verdict. I felt as if they expected me to be voluble, firing off one-liners, and I couldn't deliver. For his part Roger was easy, without any self-consciousness, choosing to talk to those with whom he could start in midsentence. If he seemed a trifle subdued compared to before the war, it was brushed aside by the pleasure people took in seeing him again. Indeed, a friend who was outside the secret recalls that Roger seemed fully recovered that night, such that the friend dismissed any lingering worry over AIDS.

I tended to be happiest on the way home from such events, rather than at them. I'd be so thrilled that we'd brought it off, and then it was such a comfort to talk to Rog about whom we'd seen. Whenever Cesar was down, we'd always say we couldn't wait for parties we gave to be over. At midnight Cesar would murmur about the guests who had settled in: "Don't they understand we have to *analyze* all of this?"

On Tuesday the sixteenth I had an appointment with my dentist in Santa Monica. I left Roger at home, having convinced him to work mornings in the study, then have lunch and a rest before going in to the office. I didn't want to go out at all when he was home, and the whole idea of dentistry seemed as vain now as rhinoplasty. I don't know if I thought the next part through in advance, but I don't think so. I was sitting in the hygienist's chair, and this perky round woman came in and prepped her table for the cleaning. I said, "I think you should know that I've been exposed to the AIDS virus."

She backed away in abject horror and ran from the room. There was a spate of frantic whispering in the hall, and then the receptionist came in and asked me to move to the main examining room. After a minute Dr. Kurtzman himself came in, obviously shaken but rigorously professional. He expressed genuine sympathy that I had two friends who were down with AIDS, then began the process, awkward for him as much as me, of donning gloves and mask. I couldn't tell him one of the two was Roger, who'd been his patient longer than I and took far better care of his teeth.

This was three months before dentists in L.A. reached a consensus that they had to protect themselves generally, because who knew who was gay or sharing needles. I should have been grateful for Kurtzman's conscientious handling of the situation, but at the time I was staring up at his white isolation mask, which seemed to clinch that I was as much on the moon as Roger. In some contorted way I wonder if the incident didn't reassure me, make it easier to believe Roger and I were in this thing together. In any case I haven't been back to the dentist since, though a molar has cracked in half. Roger wouldn't be returning either, he who never missed a six-month checkup. You get to a point where you don't seek out elective procedures anymore: if it's not busted don't fix it. Meanwhile I cried so brokenly driving home from Santa Monica that I had to pull over till the storm passed.

I had an abrupt call mid-April from Joel, to say that he and Leo were moving to New Mexico. They'd had enough of L.A., and besides, Leo was on permanent disability from the Feds and didn't have to work anymore. I was at a point where I'd fly off the handle at anything. Joel's call left me seething for days, because it didn't seem fair that he and Leo should get a paid retirement while Roger and I had to keep on working. Which didn't make a lot of sense, since Rog wanted to work and was exultant to be back. I was the one who'd had enough, who longed sometimes to cash in and do nothing, waiting for dusk on the final mountain. When I hung up from Joel I felt an enormous relief that I wouldn't have to talk to him anymore. It was still an insupportable business to hear about someone else's AIDS and have to pretend to be well and strong.

We had a bad scare in April. One morning we were up early to go over to UCLA for a checkup with Dennis Cope—I think we saw him every other week at first. Roger came out of the bathroom and said he'd had a moment of temporary blindness in one eye, almost as if a bright light had been flashed in it. We'd only heard an anecdote or two about the blindness that came in the wake of cytomegalovirus retinitis, the ravaging eye infection everyone trembled at the mention of. We knew it by its initials—CMV—and though the virus can invade any number of organs, including the brain, if you said CMV to a gay man then, you were talking about going blind. There was no treatment for it in '85.

Cope reassured us a couple of hours later that it was probably just a sudden shift in blood pressure, not significant otherwise, but he made an appointment for Roger to see Allan Kreiger, an eye surgeon at Jules Stein. Kreiger examined Roger in turn and found nothing except what are called "cotton-wool patches" in the retina. These often appeared in AIDS patients who'd come through a bout of pneumocystis, but as far as the doctors could determine, the patches remained inert. Once we had got past the crisis, I pictured the patches like puffs of cloud in the retinal sky. But I'd held my breath, heart hammering, when Kreiger focused his lens to peer into Roger's eyes.

Kreiger is as temperamentally unobtrusive as Cope, a spare eloquence underlying his measured tone. I think it was the cello he played in his extracurricular hours. Roger, who'd played the viola in high school, responded to his refined and cerebral nature. They also had a mutual friend, with political views to the right of Genghis Khan, and they traded a laugh over him. A doctor's appointment can become quite giddy when the news is good. It's as if for a moment you are no longer exiled—no longer one of *them*. The day we first met Kreiger we were as sharp-eyed as anyone, 20/20 and holding.

I don't recall that Roger had a problem keeping weight on after the first hospitalization. The Pentamidine may have quashed his appetite for a week or so, but I soon had the satisfaction of beefing him up. Regular nourishing, fattening meals are the best you can do, day to day, for a person with AIDS. No wonder the macrobiotic route is traveled with such elaborate attention to perfection. One rich man I know has a macro cook who comes to the house, and another has a box of macro delivered once a day by a Zen master.

I had my own obsessive ideas about food. I'd always been a closet vitamin freak, and for twenty years have begun the day with monastic grayness, downing a drink of soy meal and brewer's yeast in buttermilk. It tastes as bad as it sounds, and its austerity has never kept me from eating french fries and chocolate with abandon the rest of the day. I'd never managed to convince Roger to go the soy-and-yeast route. He ate heartily whenever he thought to and loved a full refrigerator, but he also had a graduate student's absentmindedness about meals. Especially for lunch, he often made do with a couple of dollars' worth of deli.

Now at last I got him to drink this glop of mine and scarf five or six vitamin pills. The totem vitamin at the time was lysine, an amino acid said to have some never-quantified effect on the immune system. Also zinc, which was severely depleted in people with AIDS. There would be times ahead when food itself seemed to become the battleground, and every wrongheaded frustration came to a head over a plate of eggs. But for now we had it easy, and Roger didn't mind indulging my steady stream of dietary hints and heaping bowls to go with them.

I was cooking in five nights a week, producing a string of fortifying meals that recalled Sheldon's pithy list of three months before, when losing a few pounds was cause for celebration. It turns out a home-cooked meal offers a double dose of magic. At the same time you're making somebody strong again—*eat, eat*—you are providing an anchor and a forum for the everyday. To the young and the impatient, dinner is something to be wolfed on the run, and the deadliest bourgeois picture of all is the same people in their same places at table.

Once you are beaten up or down by life, you begin to see why everyone longs for the dinner break, hungry or not, from Mother Teresa to men in trenches. Dinner is one of the very last corners, even in places like 1028.

We quickly turned away from anything too lean and rarefied. The finicky princess-and-pea meals favored by the foodies were suddenly insubstantial, and we avoided the grazing restaurants in favor of the all-American. But it wasn't nationalism, or even reverse snobbery. What we needed now was sheer bulk: spaghetti, baked potato, chicken with corn bread dressing. Then after supper I'd sweep us out of the house to go get ice cream. We had ceased to count cholesterol. That is a hobby for people who are in for the long haul.

Cesar, whose major dietary concession was to cut back on his beloved coffee, called to report on the trip to Uruguay. He'd discovered for the first time that he, the penniless schoolteacher, could take eight people to dinner for thirty dollars in a country that was upside down from inflation. Meanwhile his mother kept assuring him he would be fine, and to prove it she picked a basket of quinces off a tree in the yard and stewed them up. She accompanied this with a long, rambling lecture on the fortifying power of the fiber of quince, which took care of all things viral. Cesar said he felt like writing an article for the *New England Journal of Medicine* on "The Properties of Quince."

One afternoon in April, I remember, I accompanied Roger to lunch at Ben Frank's, a coffee shop on the Strip, styled in the kidney-counter spaceship mode of the fifties. I nursed a cup of tea while he had a sandwich, then I coaxed him into dessert. But I had to leave him there to make a lunch meeting with one of the *Manicurist* principals, a hundred yards down the street at Sunset Plaza, where three high-toned cafés are tucked in among the killer boutiques. L.A. is a city of ludicrous contrasts on practically every block (what else is new?), but from Ben Frank's to Pasta Etc. is still a long hundred yards.

In the old days I would have been invigorated by the mad juxtaposition, but I sat at an outdoor table that day fragile as glass, listening to this movie jerk gargle about his current diet and other phantom problems. I didn't want to be there at all; I wanted to be back at the coffee shop watching Roger finish his pie. This part is not just AIDS, I know. People still have to do business when people they love better than life are sick, and they taste ashes in their pasta just as I did. None of us has a lot to say about the current shade of peppers.

Food was gauged now by its weight and mass, and otherwise all it promised was to let us break bread together. The weight of course becomes an obsession all its own. I recall Cesar telling me of the poignant weigh-ins at San Francisco General, when men who had aged decades in a year of AIDS, so frail they could barely stand on the scale, could still summon a cheer if the pound weight hadn't dropped. We'd broken our own good scale some years before, trying to weigh Puck, with Roger hugging the squirming dog in his arms. We were left with a wildly fluctuating dime-store model, which drove us crazy with readings plus or minus four. One Saturday morning I went down to Koontz and bought a proper digital scale, and we laughed at our own excitement as I pulled it out of the box. I made Roger strip down before he got on: 141. More than a new toy, it was a way of proving every day that things were stable. He'd regained the lost three pounds, and as long as we kept him at 141 we were winning.

Robert Frost: "Weep for what little things could make them glad."

Just before leaving the hospital, Roger had asked Cope how long a person could expect to go between infections. It varied a great deal from case to case, we were told: "Two months, three months, six months . . ." I longed for him to go on and fill the ellipsis and give us longer, but in fact it was all guesswork. I remember bargaining in my head when he said it, hungry for the whole six months. He went on to reassure us that the most important element was how treatable the infections were, with progress being made all the time. But time itself becomes so pregnant when it's hardened into a number that the downward spiral of infection ceases to be real. Time is the number you fight—two, three, six, all these laughable integers. Where had the tens and hundreds gone?

Turning back over the calendar, I find a crossed-out notation for April 11: *Leave for Cairo.* There is so much left behind, I know, in the line that strikes out Egypt. But on the day in question I wrote it off as indifferently as lunch with a marginal friend, the kind that is rescheduled over and over till it disappears. The new calendar had no Cairo, its only real appointments those that drew us closer to treatment. As time is money in business, time was medicine now.

We learned there was a final hurdle before Roger could be accepted in the suramin program. A viral culture would measure the blood level of Human Immunodeficiency Virus (HIV) in the blood. Some AIDS patients had been so exhausted by the disease, had progressed so far, that the virus no longer had any T cells left to kill. It had either vanished entirely or sunk so deep in the DNA that it could no longer be traced. If Roger had no sign of virus in his blood at all, an antiviral would be of no use. Thus began a waiting game as the culture slowly grew over several weeks.

When I took this test myself in July of '86, the viral culture still had a vampire edge. They draw the blood and rush it over warm to the lab, so whatever is alive will seize the culture medium. The test costs six hundred dollars, and since the insurance company doesn't consider it reasonable or necessary, I'm still trying to get reimbursed for it. The results proved to be a dazzle of euphemism: "borderline active." What a floating border that must be, like the icy line in the Bering Strait that separates Russia from the last Aleutian. At that time I was advised to start an antiviral of my own, and by then there was something to take pre-ARC. Just fifteen months between Roger's beginning suramin treatment and me on ribavirin. Now we know that stride could have been made in '82 or '83 if the government hadn't been playing ostrich. Spilled milk, people tell me; you can't undo the past. But can't we measure the spill?

Roger and I were blessed in the friends who knew the truth. Richard

Ide and the Perloffs checked in regularly, encouraging us in the chase for treatment. They had us over and talked about it as much as we needed to, no more. None of them ever spoke a fearful word about the nightmare possibilities. They were willing to go step by step, taking their cue from Roger, and they understood perhaps better than I how much he didn't want to be seen as a sick person. Obviously it was more than his law practice that held the circle small. He didn't want to bring anyone down. Rather, after the battle of the first infection, he seemed to call his friends back to life as it was before the illness. Whenever we were together with one or another, I'd swim as best I could with the flow of good fellowship, saving my madness for telephone calls to my brother and Craig.

I think I would have been able to keep Star in the dark if she hadn't turned up in L.A. But her father died suddenly on April 15, and she had to fly to Honolulu for the funeral. When I talked to her in Hawaii, she said she'd like to stop off and see us on the way back to New York. As soon as I said yes, I knew I would have to tell her. Roger didn't want me to, and at the time I couldn't understand why. Not including how much I trusted her, how could Star in New York hurt his practice out here? It's only lately that I've begun to understand the other reasons: his pride, tenacity, modesty, all of them kinds of denial perhaps, but one mustn't forget that some are virtues. I comb back over these motives now with fine teeth, because I want to be as strong as Roger, as full of life. And stoic has not thus far been my strong suit.

At last he agreed that Star could know, but he was much more concerned the day she arrived that we take her out for lunch and let her talk about her father. Only after she'd told the whole story did Star and I go off alone. We took the dog for a run in the park at the top of Laurel Canyon. I remember sitting on the grass and crying as I told her, not two minutes after she'd enthused about how terrific Roger looked. Star knew me for eight years before I met Roger and always said the relationship burned off the misery of my youth and freed me to write. She understood the stakes here better than anyone.

Star told Roger she appreciated what an act of trust it was to let her in. As she pledged support, I could see how shy it made him. That night the three of us went to a party at Susan and Robbert's studio in Inglewood, full of art folk whom we didn't feel the need to impress and who didn't in any case take much notice of us, since we weren't buying. Pam Berg, a merry and quick-witted Englishwoman who's head of the Graphic Arts Council at the County, asked me how Cesar was doing. Then she spoke of a colleague of hers who was working for the AIDS Project, in the buddy program. His person with AIDS had been doing fine, but now he'd had his second bout of PCP, and matters looked dire. I recall wanting to run from the details, yet had to have the full picture first: how old, sick how long, how early had they caught it, what treatment?

When we arrived home at midnight, I asked Star to finish a roll of pictures in our camera. There are eight shots of us on the sofa in the living room, with Puck barreling into a couple of them. These were the first since January, and I know it was in my head that we must have further evidence of us, though I tried not to think why. Roger's hair is grayer and thinner here than it was in Greece. His cheeks are slightly drawn, his color still pallid—not exactly what the French call a *coup de vieux*, but he'd aged five years. Yet, especially where he's smiling, his eyes are bright and the play of his smile vivid as ever. My own rictus of a smile is pretty frozen, all brave front. In one I'm looking at Rog as he smiles at the camera, which throws me back to a Sunday that spring when we went to see *Purple Rose of Cairo*. All through the movie, every ten seconds I'd turn and stare at Roger's profile in the dark, drinking him in, my ears acute as a wolf's as I listened for people coughing around us.

I see now we were back to the rhythm of weekends, making forays off the moon. One Sunday at the end of April we drove out to Malibu, to the Getty Museum. For once I'd actually made a reservation, instead of lying my way past the guard at the gatehouse. We'd watched the Getty extend its collections masterpiece by masterpiece ever since the museum opened, and we both found the Roman villa setting as bracing as an "E" ride at Disneyland. After we returned from Greece we promised ourselves we'd spend more time at the beach, and one perfect weekday we'd played hooky at the Getty, drifting through the antiquities, our eyes still dazzled by the white light of the real thing. In the Greek gallery we lingered over the fragment of a marble relief, a wounded warrior and his comrade, pure as a line of the *Iliad*.

Now, in April, we headed for the room upstairs that houses the illuminated manuscripts. A whole truckload of them had recently been acquired in Europe in a single bullion sweep. The Ludwig collection is kept in theatrical darkness, the only light within the cases, playing on the open books. We bent and peered in each jewel box, delighting in the microscopic artistry that adorned these gilded medieval windows. I always made sure no one was standing at a case before we approached, because I didn't want anyone breathing on us.

Craig arrived next day and stayed three nights. He'd visited friends in San Francisco and bought a dogleg ticket so he could see us too. He wasn't afraid to say it out loud: "I have to come now; I don't know how many more times we'll see each other." He knew the ground rules coming in. Because Roger had no idea that Craig knew of his diagnosis, it would have to be a two-tiered visit. If Craig was a little jarred not to be able to share the equal situation with Rog, he also understood the difference. He hadn't been through the full blast of an infection, and Roger had. In truth, with only three small lesions, Craig was suffering the metaphysical side much more than the physical. So in that way he was more equally matched with me than with Roger, though there I go with one of those comparisons that get you shot at sunup.

When the three of us were together, Craig and I let Roger set the tone, though I recall Craig talking about AIDS quite freely, given the strictures. Roger didn't avoid the issue at all, encouraging Craig to speak of it exactly as he had encouraged Star, by his willingness to listen and not interrupt. I suspect such openness in Craig and Cesar helped him to feel less alone as well, even if he was the silent brother. Still, it was a curiously schizoid few days—talking with Craig about nothing but AIDS whenever he and I were alone, piecing together every anecdote we knew, trying to figure patterns that hadn't been reported yet. Then Roger would come home from work, and we'd shift gears and cook dinner. Late at night after Roger went to bed, I'd sit upstairs in the attic bedroom, strategizing once again with Craig.

The Perloffs had asked us out to the ballet, a traveling company at the Wiltern Theater. The dancing was pretty overripe, and I mostly felt a sense of vast irrelevance at being there. How was it the world went on like this? Roger enjoyed himself with the Perloffs, but Craig and I were like two anti-intellectual, fidgeting children. It reminded me of Boston ten years before, when Craig and I were sometimes as raucous as fraternity mates—delayed youth, since neither of us had had a gay friend in college. Roger had been indulgent then about Craig and me at our noisiest, and he seemed the same way now, probably glad to see me laughing, however black the joke.

We came roaring home from the ballet, stopped off for quart-sized hot fudge sundaes and brought them home for a feed. By then the three of us were rollicking. There was a hilarious rush of pleasure at the prospect of immediate gratification. We joked about how quick it would be over, even as we gorged the ice cream. I ask Craig what else he remembers from that visit. The next morning, he says, when he had to get up early to catch his plane, he came downstairs half asleep. Roger was already "humming around the kitchen," getting ready for work. I keep playing that humming moment over in my mind, for I was still on the old Jean Harlow schedule, fast asleep. I love recovering any unguarded moment I might have missed, especially from a good day. It tells me all I need to know about how Rog was doing.

Craig's visit also slots it in my mind for a certainty that Rog had moved back into the front bedroom—*our* room. Craig could have taken the room by the pool and needn't have been exiled to the attic. But I wanted to give Roger all the space he needed, and partly too—irrationally—to separate Craig and his New York microbes. As a child I used to fantasize late at night that my bed was a raft set loose in a shipwreck. Of the many small victories of being restored to life, sharing a bed again felt like a real turn in the war. The deepest habit of normalcy is unconscious, as one of us would turn in sleep and hold the other, spoon fashion. This was the point of maximum stillness, proof that things were right again. Roger always averred that he'd invented the spoon on Sacramento Street after we met in '74, in order to ease my birdlike hyper nerves so I would fall asleep. Thus the middle of the night is when I feel the loneliness most. Even half-unconscious, I still turn and tuck in a spoon, preferring the memory trace to nothing.

We made love again in April too. I have to speak of this a little, though I feel the tug of my right to protect it, because I live in a generation of gay men from whom Eros has mostly fled. It's true that in time the holding close and spooning were more the ground of love than passion was. That is, we grew to need the repose of each other more than the heightened intensity. Roger was often too unwell and I too strung out to think of naked sex. But the burrowed place of holding on, where life was the same as ever, still could release an exhilaration that gathered to a peak. The first time Roger came after his hospitalization, he was almost crying even as he gasped with pleasure. England and the ballet might be over with, but not us.

Seven years into the calamity, too many gay men have lost the will to love. The enemies of our people—fundamentalists of every stripe, totalitarians left and right—have all been allowed the full range of their twitching bigotry. Though gay men have begun to understand it is something in themselves these upright men so fear, too many of us have internalized their self-hatred as shame. That the flesh and the spirit are one in love is none of the business of the celibate men of God, especially those who believe they rule the province of love. But the mission of the homophobe is more pernicious even than his morality. He wants every one of us to be all alone, never to find the beloved friend.

A man ought to be free to find his reason. Not that freedom alone will serve it up: it requires the gods' own fury of luck to get two people to meet. But when it finally happens, two men in love can't rejoice out loud—joy of the very thing everyone burns for—without bracing for the rant of prophets, the schoolyard bully, and Rome's "intrinsic evil." I try to remember that we fight as a ragged people to outlast the calamity so that others can sleep as safe as my friend and I, like a raft in the tempest.

Random memories of that April are spots of time. I can see Roger sitting at the desk in the study, writing a condolence to Star's mother, while I sit watching from across the room with nothing to do, till I run out and get him a milk shake. When in doubt, feed. I remember a brutal op-ed piece in the L.A. *Times*, written by a doctor at San Francisco General, where he tried to make graphic just how desperate the disease was. Yes, fifty percent of those stricken were still alive, but only because they weren't dead yet. They would die soon enough. It's one of the times I recall Roger crying, and I own the idiocy of my response. I said, "Look, Rog, the worst that can happen is both of us will die," which *really* set him off. I try to remember that bit of backfire when someone says something off the wall to me.

I remember renting *The Night of the Shooting Stars* and both of us being enthralled by it. A lucky draw, since I was a wash at the video store, where I would dither among the *Rambo*s and be unable to locate anything. My brother had urged me for years to see *The Horse's Mouth*, with Alec Guinness. We laughed all the way through—that great moment when Guinness disguises his voice on the phone, declaring he's the Duchess of Blackpool. Yet when it ended and the madly sane painter sailed away down the Thames, I started to sob uncontrollably: "I always thought I was going to be Virginia

Woolf, and now look at me!" A comic enough disguise all by itself, if it weren't so sad.

Gottlieb managed to throw us both off balance one morning after an examination, with a soft-spoken exhortation: "Make every day count." Roger and I groaned over that one, comparing it to the moment in *Great Expectations* when Miss Havisham snarls at the terrified Pip: "Play!" Sam tried to place the doctor's remark in context. It didn't mean we were supposed to stumble around looking for roses to smell; rather, we had to learn to savor life as we used to—except now it could only be day by day, whether we liked it or not. I was acting as if I could keep Roger's situation stable only by staying depressed and morbid. If we laughed too loud, the Big Foot would stomp us. My guilt and doomsday magic were keeping us from reclaiming the fullest measure of life, and if it kept up, I would rob us after all of the love we had. The question I needed to face was: Where had Paul Monette disappeared to?

> *The only answer I can come up with is that the last and best and only PM I care about is the Paul of Paul & Roger, and now that I live with the sword over my head there is this constant pain and stupefied disbelief.*

But don't you believe you can help him to live, demanded Sam. This "Paul & Roger" that I believed in so fervently required of me sometimes that I be stronger than I was. The loneliness of the secret was turning me inside out, and gradually I would have to accommodate more people like Star and Craig. I had to build myself a support system with others so I could be the mainstay to Rog. Some mornings I would lie in bed swirling with horrors, unable to get up at all. Fight to seize the day, Sam said, call somebody even worse off than you are, *engage.*

Roger after all was busy at work, husbanding his energy, brimful with a world that had welcomed him back. He relished coming home to me—to us—more than ever, but now too often he'd find me glazed with pain and emptiness. Even to me it was sounding too close to what happened during the bad months of '81, when I was reeling from my obsession over Joel, unable to work. Back then, Roger had finally admitted it was getting to be an ordeal to come home to my misery every day. Not in so many words, now he was pleading with me to get the AIDS despair behind me. *I* was the one who hadn't recovered yet.

I hope I make it clear that both were happening at once: raft and tempest, peak and valley, the prostrate Harlow and the ice cream messenger, all in a day. So much was Roger not blaming me for his illness, or beating me up with his own pain, that I'd fashioned a way of blaming myself twice over, till I was more invested in self-punishment than in relief. Sam told me I was fixated on the unfairness of it all, refusing to cope with this thing that didn't belong in our lives. Sam practically shouted at me: There is no fairness! I was clinging to the iceberg instead of the raft. Life was about survival and challenge—so *meet* it.

The suramin ticket came through at last. We went in for the first dose on Friday, May 24, with very little sense of what to expect. We knew that Roger would have to be officially admitted to the hospital and pay for a half-day's room, even though he would be in the Clinical Research Center only three hours every Friday. Two other patients had already begun the program during the previous week, but on that first Friday Roger was the only one. We'd been told it would be administered intravenously, and we had the name of the doctor in charge, but all we knew for certain was that this was the only game in town.

Clinical Research was altogether different from the tenth floor or any other bustling space at the medical center. Here the corridor was quiet as a Zurich clinic, with three or four rooms on each side, only a couple of them occupied at any given time. Cancer patients mostly, I suppose, watching daytime television while something quixotic dripped in their arms, and maybe a wife or husband flipping through a magazine. The room at the end of the hall on the right was a two-bedder, which happened to look out into the thick crown of a banyan tree. "It feels," I wrote in my journal, "like we're in a tropical hospital somewhere, getting treated for a rare jungle infection." A dietitian came in and took an order for Roger's lunch, though it was only 9 A.M. I don't know why that seemed so generous—we were paying a hundred dollars an hour to be there—but this friendly woman was very proprietary about feeding her guinea-pig patients. The chief cook in me applauded the sentiment mightily.

Then Peter Wolfe came in. *This* was the clinical researcher? He was disarmingly young, with a shag of blond hair that slumped across his forehead and gave him rather a hooded look, though he tended to peer through the thatch with wry amusement. It happened that he'd gone to Harvard, and Roger flushed with pleasure as they traded memories of Cambridge. It turned out Peter had also read a novel of mine, which was proof enough to me that he was gay, but on second thought he was pretty forthcoming with that information himself. Explaining that he had been treating AIDS patients since his first day as a doctor, he spoke simply and feelingly of looking down at a stricken man in bed and thinking: "This is me."

As he went through the technical rigmarole of the drug, dosage and possible side effects, Roger and I had much the same reaction. We were in the care of one of our own, and here I don't simply mean gay. We shared with Peter a common geography of the mind, which only by chance happened to have the Charles River meandering through it. Still barely thirty, Peter was the age of the men and women Roger had worked with as senior tutor at Dudley House. Peter also had an eccentric sense of humor and love of puns that mirrored Roger's own, to which Roger responded with quick delight.

So began the first elixir. I soon came to think of that room looking out on the banyan tree as a safe haven. Over the course of the summer it grew populous with its own queer and hardscrabble island folk.

Later, of course, when the Swiss Family treehouse came crashing down, the island seemed a mirage, but now the stronger memory is how happy we were to beach there. We still froze with terror at every bruise. Only a week before, we'd almost missed a wedding in Venice when I found a purple blotch on my upper arm, and now I searched both of us head to toe every day. But from the moment of that first dose in the tropics of the CRC, we had an ace in the hole. The picture in my mind was of the virus pulling back, shrinking into itself as the blood woke up.

I can't pretend the panic and despair shrank in direct proportion, but we left Clinical Research that first day with an awed feeling of gratitude. Roger had been required to sign a consent form, and I had witnessed it just below. That our names were twinned there meant more to me then than the marriage license the laws denied us. Peter Wolfe read the form aloud, and there was a sentence to the effect that one would have the satisfaction of knowing one was advancing the search for a cure. I sat up straight at that, quick as an old soldier to salute. Hope is not the same thing as believing anything higher, but it suffices unto itself. We had two secrets now, the illness and the treatment, and the second was so extravagant we fairly glowed. It's a long way still to being a hero, but within a week the teary boy who would never be Virginia Woolf finally admitted there was something to say besides "More dead."

> *. . . frozen too long, and there is this ache like tears that wants to burst. It's like I died, and I* didn't *die. We are here, and we love each other, and now I have to find some work. Sentence by sentence, nothing by nothing, even if I can't sing. Then hum a few bars at least. Whistle a bit in the dark. We cannot all go down to defeat and darkness, we have to say we have been here.*

# ·VI·

Evenings at the brink of summer are yellow gold across the city's western face, as the sun narrows toward the ocean, eye to eye with the white buildings of the coastal basin. The setting sun is especially prized in late May and June, because the Catalina eddy hugs the city often until midafternoon. Then clearness seizes the landscape. Summer is something else again, sunny all day long, till the light expires of heat and boredom after Labor Day. Perhaps because we had come from winter places, we were finely tuned to the threshold effects of summer. In any case, Memorial Day usually found us thrashing in the garden, giving a nudge to the rattling pool heater and pulling together a barbecue.

This year we spent the weekend lying low, alert to any side effects of Dose 1, but they never materialized. So Monday evening we asked Dell Steadman over for supper, as if we couldn't let the holiday expire without a show of colors. Because Dell was both doctor and friend, we'd talked with him extensively in the weeks before the diagnosis, but since he wasn't on the short list of those who knew the truth, contact had lately been minimal. Thus Dell was in the peculiar position of suspecting Roger had AIDS yet having to leave the matter unspoken, and without any pregnant pauses either. He was more than up to the challenge, however, regaling us over dinner with his Late Roman view of the tattered state of the world. No doubt we were ready to laugh, fortified as we were by the quaff of elixir three days before. I know it felt good just having somebody in again, so much so that the wide swath of summer began to tantalize. For the first time I thought there might actually *be* a summer.

Two calls came back-to-back that night. The first was from a business acquaintance in New York, who started to gossip about Bruce Weintraub having AIDS, as if I were party to the knowledge already. I put him on fast rewind, and he dryly observed that Bruce had been hospitalized a week with "regular" pneumonia. I felt two furies at once: protective about our own secret and angry at the slur on Bruce's privacy, as if to try to escape the rumor mill were an act of contempt. But even the anger couldn't cover the queer sickening feeling I've had fifteen different times in the last three years. How could Bruce be sick? You never stop asking that. There's a strange recurrent wish to believe the epidemic has claimed enough, even as the shock waves widen. Above 8.5, an earthquake is said to liquefy the earth. I recalled joking with Bruce about AIDS in front of the gym two months before, ridiculing the hysteria of the very man who was calling me now.

I put down the phone and it rang again. It was one of the *Manicurist* producers, typical of the breed that hates all holidays because of the three-day lag in deals. By now I had written perhaps two thirds of the Whoopi screenplay, with rabid input from the studio along the way, though only a couple of people were privy to the actual pages. Pages are not required, however, by those whose drug is opinions. "But has he *read* it?" I asked a few weeks later about a particularly wacko criticism, only to be told in oracular tones, "David doesn't read, he hears."

The producer himself had in fact read pages, and after a fulsome wheeze of praise he explained that the studio was basing its decisions on how well a script conformed to a certain grid, which was all the rage in screenwriting courses. The great sage who deduced this foolproof method had codified every hit movie of the last twenty years and figured out that the main "plot point" always occurred between pages 26 and 28, with a secondary plot point around page 90. Therefore, said the producer, it was time to deconstruct my script and nail the plot on 26.

The echo effect of these two calls resonated over the next month like force and counterforce, demanding what little energy I had left after me and Roger. It was probably a foregone conclusion which would dominate, and now I see that I needed to reach out to Bruce even as I washed my hands of comedy, though I didn't know any of that on Memorial Day. For the present I was merely shaken by the calls, though accustomed now to the phone ringing like a siren.

By Friday, when Roger received Dose 2, we'd been alerted that there might be a reaction, since the other two suramin patients were feeling feverish and wilted. But Rog was strong enough to go to work for the afternoon, and we decided to fill in the weekend, half full anyway. An old friend was in town from Toronto and eager to see us. Bohemian to the core, Gordon had improbably been appointed director of the Canadian Book Council. Before his elevation he'd lived in L.A. for several years without a green card, and been instrumental in opening A Different Light, the local flagship gay and lesbian bookstore. Since Roger and I last saw him, he'd become almost laughably respectable, commanding offices in three provincial capitals.

Gordon arrived at sunset, with irises and champagne. I had determined not to bog down the evening with my oppressiveness and to minimize any talk of Roger's recent illness. Gordon himself had had a bout of shingles in April, a perfectly respectable disease in general circulation, except there were far too many cases of it among gay men who went on to develop AIDS. Craig had had shingles; the power broker who got us on suramin had them; Roger would have them. I saw a man at the gym last month and laughed hollowly as he related with antic dismay that he'd just had chicken pox, the childhood variant of the shingles virus. Everything is in clusters now. What is innocent as the sniffles in single cases grows specter-thin with terror in groups. Needless to say, there was reason to think that Gordon would be glad to speak softly of illnesses that were nothing to worry about, nothing at all really.

It was great fun to listen to Gordon's tales of cultural czarism, and we did manage to keep the conversation virus-free. But I also couldn't hide my jangled state and threw it all onto my work. I even asked at one point, "Gordon, what should I write about?" Thinking as I said it of the elderly Tennyson, begging his wife and children to slip ideas for poems under the door of his study because he was all written out. Gordon replied without a pause: "Write about what's happened to desire."

Next day Gordon dropped by with his friend Anne, and we all had tea by the pool. Anne had recently been through kidney surgery and nearly died. I remember staring at her as she looked peacefully at the flowers, trying to figure how she'd stood it alone, always a sense of kinship now with anyone who'd been through fire. When the two of them left, Roger took a soak in the tub, a little feverish and washed out from the drug. He studied his hand for a moment and said tearfully, "I guess if anything happens to me, this should go to my brother."

The ring was a sapphire set in white gold that his father had won in a card game decades ago. I was there the morning in '75, at the old apartment on Chestnut Street, when Al slipped it off and gave it to him. At the time I was jarred by the flash of it, a shade too Damon Runyon for Rog, but over the years I'd come to see it as one with the gentleness of his hand. I told him he mustn't think that way, now that we finally had some hope, but he went on to wonder aloud if he should keep working at all. "Maybe I don't want to," he said with a weary sigh. I swore he'd feel himself again as soon as the weak spell passed. It didn't somehow factor in that I'd spent weeks of my own wondering why I kept plodding away at the computer. Did he really mean to consider cashing in? It was the only time he ever wavered about work, so I hope I didn't too hastily close the subject off. Yet the deal between us always permitted the reopening of anything, or how would we ever have gotten to California, or Roger into private practice?

By the next weekend we'd instituted a regular Sunday dinner on the front terrace, with four or five around the glass-top table: the first time I'd laid out a table or baked since Christmas—a lemon cake, I think it was. "No matter what happens," Cesar used to wag when the Christmas cooking was in overdrive, "Mrs. Ramsay gets that leg of lamb on the table." Sunday nights were the serial version of the silent film, where if you didn't know what was racing in my head, the terrace on North Kings Road was casual as ever. Those evenings surface now like a string of summer islands.

I note in my journal that Roger processed Dose 3 so well that he swam fifty-two laps of the pool—only a seven-stroke pool, but who's counting? I also note that a main subject under discussion now was whether or not I ought to be on antidepressants. I'd just had a strained session with the Ferrari doctor, where he heard the state of my head and leaped for the phone, eager to refer me to a psychopharmacologist. His considered opinion seemed to be that I would love antidepressants. The shrink in question had apparently brought Ferrari himself up to full potential. I threw a damper on his enthusiasm when I told him my Writers Guild insurance paid only fifteen dollars per psychiatric visit. Otherwise shrinks could bankrupt the guild in a matter of months. Since the psychopharmacologist was a hundred and fifty a visit, Ferrari shrugged me out with a prescription for more Halcion. What was I doing still seeing this man? It was almost a kind of paralysis, as if I didn't deserve any better than his indifference. He had seemed a perfectly adequate doctor before the war, when nothing ever went wrong and all of us were going to live forever.

Joe Perloff advised against antidepressants if I could make do without them. Sam and my brother concurred. There was a strange curl of vanity here that kept me medication-free. I knew a couple of AIDS-related people floating on Xanax and Sinequan, how they ballooned with weight—their faces round and bewildered as babies, like Lennie in *Of Mice and Men*. No, thanks.

With Dose 3, the cast in Clinical Research grew. A certain Mr. Appleton appeared for his first dose, his encyclopedic knowledge of the antiviral territory dwarfing my own. He also seemed in demonic good health, brisk and alarmingly chatty, though Roger and I came to enjoy his tirelessness. He'd found out his T-cell ratio was reversed—I don't think he had any other symptoms—and talked his way into the suramin program by sheer force of will and a thousand cascading phone calls. Appleton always seemed a fine example to me that one didn't need higher contacts at all. A murderous push and refusal to take no for an answer had got him where he was. He had a home-brew recipe for HPA-23 that he'd got off a biochemist, in case the suramin didn't pan out. Dr. Wolfe blanched a little at his torrent of questions, but nothing daunted Appleton. He had that quality of utter belief in his own story, like Ishmael, and a sense of being accountable only to himself.

It was that day, I think, that Peter Wolfe happened to glance at Roger's hands and remarked that the moons had disappeared from the nails. It was a curious minor feature of the disease, he said, and didn't seem to mean anything, but I recall being jarred by the whole idea. The setting of the moons had some kind of inner planetary echo about it, indicating how very subtle the virus was, casting its shadow in places that had no pain or symptoms, no reason at all except to be bizarre. Similarly the intensification of dandruff, which now required a brown shampoo of industrial strength. The most casual things took a twist, as if to remind you that nothing in the body was to be taken for granted anymore. That is what aging feels like, isn't it? It's common among gay men now to say we're all eighty years old, our friends dying off like Florida pensioners.

Somewhere along in there, Cesar flew down for a few days. After nearly two years of me buoying him along, however manically, it was he this time who throbbed with life. He kept reassuring me how well Roger looked, and gallantly dismissed his own recent struggles with the illness. His hair was noticeably thinned by the chemo, his leg still swollen and suppurating, but he was irrepressible. "Don't worry," he said, "I'm eating a lot of quince." It seemed important to him to get it across that all of it could be borne and processed. Nothing of life was irretrievable: that was the unspoken promise.

One gray afternoon the three of us went to the old Doheny estate in Beverly Hills, empty now that the American Film Institute had vacated it. We wandered up and down through the gardens, the latter a bit brambled around the edges. We talked and laughed so comfortably that the afternoon has grown seamless, green as the gardens of a dozen years. Roger and I told Cesar about the lone Doheny heir in the thirties, a young man of ambiguous despairs who fell in love with the chauffeur and, unrequited, shot himself. I don't know even now how much of the story is true, but it's part of the pulp mythos of Beverly Hills, a properly Proustian end for the scion of an oil barony.

Later we strolled around the city, and Roger went into a shoemaker's to buy a pair of laces. Though I stood outside joking with Cesar, I recall being choked with emotion at the modesty of the errand. The laces are in a drawer in the bathroom, still in their cellophane, and they evoke a certain cast of Roger's mind, his satisfaction with details. I never buy shoelaces; I throw the shoes away. And it's only a beat from there to his sensible wingtip shoes and his flat feet, and his mother telling me after Roger's funeral about giving her own mother's clothes to Goodwill. "But not her shoes," said Bernice defiantly. "No one can walk in her shoes."

What fascinated us most about Cesar during that visit was the tale of the swimmer. He'd spoken vaguely of the man for a couple of years, but only ever called him by his sport. In the beginning he was one of Cesar's private tutorial students, though I can't remember in which language anymore. Italian, it must have been, for Dennis the swimmer himself was Italian-American. About thirty years old, and got his nickname from the laps he did at the Y before his lesson. Now Cesar was talking as if they were getting rather thick.

It wasn't so different from the way he'd talked about Jerry nine months before, but here Cesar was even more enamored of the whole idea. Indeed, Dennis sounded like a catch. He knew the difference between books and their press coverage, and he and Cesar would talk away pots of coffee. It was even the sort of relationship Cesar had somehow missed: mentor-pupil, fifteen years between them, passing the history. The only problem was, he had some sort of live-in situation. I never could be sure if it was a current or former lover, only that Cesar was maddened by the complication. It also appeared that Dennis had buried two or three friends with AIDS, caring for them right up to the end.

Something in me didn't like it. Was he some sort of angel of death? Or was he out perhaps for a bit of a tease, now that the rose of his youth was past? There wasn't any sex between them, so who was playing with whom? Yet for all my reservations, I could see how it made Cesar happy, gave him a sense of identity beyond his illness. What business was it of mine whether or not it was real? Relations on the battlefield are not governed by the same rules. Real is not a function of time. Or perhaps the pitch of time itself is the passion, and makes its own reality. In any case, I was intrigued enough to throw out the notion that Cesar should bring Dennis down to L.A. for a weekend.

I don't think I'd ever seen him brighten so much: "Could I?" Though Roger and I had known a couple of the men he'd been with—three months, a year, one of them married—there had never been the occasion where we could be together as two couples. I remember the day in the fall of '74 when Roger and I drove down from Boston to Cape Cod so he could meet Cesar—like bringing him home for inspection. That was what Cesar wanted now.

Half-naked out by the pool, where he wasn't too shy to sunbathe even with his spots, Cesar told me about going to Rio after Uruguay in April. He'd been too weak to plunge off by himself, so he signed on for a day or two with a tour bus. At one point they stopped at a gorgeous beach, and Cesar stood weeping on the sand, thinking how far he was now in his hobbled condition, with his elephant leg, how far from cavorting and running in the sun. One of the tour guides came up to him and gently touched his shoulder. "Don't be sad," she said. "Come over and have some lunch." And she led him down to a picnic with the other guide and the driver, and they all talked Spanish and made him laugh. "Isn't it wonderful," he said, "how people take you in?"

One evening Roger had to go out to a Room For Theatre board meeting. He agreed that he might get overtired, but he always enjoyed the gatherings, and they valued his opinions. Besides, he was learning how to pace himself. Cope and Wolfe and the circle of friends all thought he should reach and push a little, fight for his stamina. I stayed home with Cesar and watched *Risky Business*, it having been decided that Whoopi's foil in *The Manicurist* had to be more Tom Cruise. Cesar thought the movie was a hoot; I loathed it. Mostly I paced in my head, worrying about Roger till he got home. I think it exasperated Cesar that I couldn't stop fretting. He could see that Roger was dealing with it, so why couldn't I just let him? And that tied in with an old issue that nettled Cesar, the dependence of couples.

He would usually exclude Roger and me from what he thought of as loss of self in the long-related. Ironic that we firmed it up that night: he would come down with Dennis a few weeks later, to shoot for a little interdependence.

On Saturday of Dose 4, we watched the hostage footage of TWA Flight 841 from Athens. We had taken that very plane a year ago to the day, Athens to Rome, without incident. My parents called to say we couldn't go to Europe anymore, it wasn't safe. No, of course we can't, I said, glad for any excuse to explain why we wouldn't be traveling this summer. But it wasn't a sad anniversary, not like Cesar weeping on the sand, because we were busy with treatment, cupping it like a candle in the wind. No side effects except for a temperature on Sunday.

Next day Richard Howard called out of the blue, from Houston. He'd been a kind of mentor to me when I was first writing poems, and his dramatic monologues had stirred me deeply. He and Roger shared a passion for French, of which Richard was the primo translator into English. But more than anything else he has the credit for introducing Roger and me, having coaxed me to the fateful dinner party on Revere Street. I'd just staggered off the train from New York after a weekend of bad karma with a friend who wanted to be a lover. "Come along, dear," Richard had said. "It'll be an early evening." Small miscalculation: the evening went on for twelve years.

In the summer of '85, Richard was teaching in Houston, a course called "How to Recognize a Modern Poem, and What to Do Till the Doctor Comes." I hadn't spoken with him in over six months, and he was startled at the blackness of my despair, which I attributed to Cesar. "But this is terrible," retorted Richard, as if there had been an affront to the natural order. "You and Roger are the sort who know what to do with happiness. I can't bear to think of either of you suffering."

Then he went on to remark that David Kalstone, a literary lion at Princeton, had been ill with pneumonia just like Roger, and was feeling much better now. I don't think Richard was being ironic. Nor did he realize how clearly he was telling me Kalstone had AIDS. He meant to do quite the opposite, surely: there *was* still a natural order, and everything wasn't dire. Regular pneumonia was such a refuge still, like a cold that leaves you no worse off than cozy in bed with tea and a novel. It's a problem having to do with my perception, but I lived too long with pneumonia as a smoke screen to be able to buy it as a true diagnosis.

I reported to Craig the nontoxicity of Dose 5, and he told me his Houston researcher had mentioned that his own four suramin patients were all sick from the drug. Not us, I thought. But right after the clinical note in my journal, the fear breaks through the optimism that counted every hour now by the number of the week's dose.

> *A dozen times a day, a hundred times, I think about Rog and what it would mean to lose him and I go to pieces inside. We had a supper delivered from J. Spector and a lovely quiet talk. But where am I going? All I want is what we have to go on and on. The world out there, I don't understand what they want, how do they bear the matter of time?*

During the Dose 5 weekend we spent the evening with Peter Wolfe and his friend Jeff. It was clearly a complicated step for Peter to take, seeing us socially, especially with such a large caseload and the need to get some distance. But it felt terrific to be appreciated as something more than Friday at nine at the CRC. We drove next day to Laguna to have lunch with Mary McMeekin, the realtor who found us the house on Kings Road and whom Roger had toured through Paris twenty-five years ago. I realized even then that Roger thought through whom he wanted to see, not by way of valedictory but rather to keep lines open to things in his life that mattered to him. Anything that harked back to Paris was rich with feeling and bathed in a golden light.

In that regard the most important letter he wrote after his diagnosis was to Madeleine: 24 Place des Vosges, an attic apartment in the old palace, a warren of rooms hung slapdash with paintings—her own, her father's, a Renoir, a Vuillard. She responded to Roger's news with an instant cable, saying a letter would follow. MILLE BAISERS, MADELEINE. During the next month three packages arrived: *pâté de foie gras*, rum babas, *marrons glacés*. The chestnuts came in a wooden box, individually wrapped in foil and tissue and redolent of a very specific Paris—chestnut trees lining the boulevards, roasted chestnuts in paper cones. All through that summer I remember Roger looking up from work or a book and announcing with satisfaction, "I think I'll have a *marron glacé*."

The Sunday of Dose 5 was a national holiday in the people's republic of West Hollywood. The kickoff to Gay Pride Week was a parade, which marked here as elsewhere the anniversary of the insurrection at the Stonewall Inn. Roger and I parked off Sunset and walked down into the checkered crowd on Santa Monica Boulevard. I worried about the germs and wouldn't let us buy any streetside food. As I watched our people mill along the sidewalks, those still young and funky carousing in groups as the tatty floats cruised by, I was too disconnected to feel much pride. Still, Roger and I reminisced about the first gay parade we'd ever marched in, '75 in Boston, holding hands as we walked across the Common. I remember the marchers that day smiling from ear to ear, but not very loud, almost speechless with joy.

An eon later we threaded our way down the jammed sidewalks of Boys' Town, with punkers and beer drinkers, queens and commoners. A couple of groups paraded with AIDS banners. At one point an open car cruised by at ten miles an hour, boisterous people sitting out on the trunk. In the front passenger's seat sat a man with AIDS, waving gamely to the crowd on the boulevard, rather like a blessing. His lesions and his hollow look filled me with terror. I didn't want Roger to see him and turned us away: it was time to leave.

As week after week went by and Roger had no bad reaction to

the medication, summer began to have the feel of stasis. By late June a solstice point is reached in southern California, such that the weather never changes from day to day: cloudless, bright hot, then at night cool and star-clear. Cesar had managed to convince me he was holding his own; and Craig was fine, just one new lesion. Thus when certain nightmares intruded now I began to feel strangely removed, as if we might be a new generation of AIDS that would just squeak through. This is not entirely an illusion. In the seventh year we have reached at least a second generation, perhaps a third, and each with a better shot at holding ground. *Living with AIDS* is a rallying cry now, and the men of '85 were the first division to hum a few bars.

One afternoon Roger said he was stopping by to see a client about signing a will. The man was a studio publicist, and he'd been in the hospital with meningitis. Roger was naive enough or sufficiently in denial not to realize he'd be walking in on an AIDS situation, but this was still before we had a clear picture of dementia and the horrors of the brain. Roger found the man bedridden, very weak and disoriented. A sister had come from the Midwest to stay with him and had engaged a nurse, but she couldn't remain away from her family any longer. It didn't sound as if it would be very long. Roger had to be a sympathetic ear to her and her brother both. Even at the best of times the signing of wills lacks a certain charm. When Roger came home and explained it all, ashen and pained, he was most upset to think that when the sister left, there wouldn't be anyone there. Was AIDS ever mentioned? Not by name, but pray, why else were forty-year-old publicists demented and dying?

On June 25 I had a session with Sam where I talked much more about the anger than the despair, about all the things that would set me off, screaming at people over the phone. At the same time I was afraid to be tested or find out what my numbers were. The AIDS antibody test was finally in place. Sam suggested that I was turning denial into a state of continuous tantrum, so defiantly did I not want the issue in my life. He also made me think about my guilt at still being healthy—even more, that at some level I felt relief that I wasn't the one with AIDS.

For me this always means that I go back and stare at myself as a lost child, with no words to explain how guilty I feel, being able to walk when my brother can't. The suppressed rage about the injustice, the astonishment that the rest of the world goes on its merry way—I'd been there before. And now I had to fight it anew and somehow make it okay that I wasn't sick, for the reality of Roger's condition was independent of my guilt and dread. My brother remarked one night, as I raked it all over with him, that sometimes I seemed to believe that if I suffered enough I wouldn't get sick and Roger wouldn't die.

Despite my contempt for my mood swings, I remember what bothered me most then was how little of them I could share with Rog. I didn't want the down times to distance us, even if I understood I mustn't burden him. I also agreed with Sam it would be good for Rog to be talking to a therapist himself, and I'd mention it every so often, but he was hard to bring around. He'd gone through a long therapy in Boston plus a year and a half of analysis. When he finished with all that, he checked out of the system. Finally he admitted it might be a good idea if he explored his feelings with a neutral ear, and I think it was the visit to the man with dementia that tipped the scale. I was the one who suggested Harry, a friend of ours who did a good deal of work with gay men and now with AIDS. He and Roger used to go to ballets and concerts together, since Harry had a lover who was immune to high culture. On their way to the Music Center one night in June, Roger asked casually about Harry's AIDS group.

Harry replied with a chill shrug: "They all said they weren't going to die, and they all died. We've gone through two groups now. We're not going to do a third."

Deflating as this was—and empty as the bravado is in any of us who profess to be toughened by having seen it all—Roger called Harry one morning and asked if he could see him professionally, just to get an idea where he might go for help. What Roger was really doing, I think, was finally admitting he wanted to tell more friends. But Harry pulled up short and said it wouldn't be proper. Of course he had every legitimate reason professionally to take the position, though that didn't mitigate my anger. And more troubling than that, Harry suddenly vanished—no more symphony, no more connection at all. Unfortunately, the disappointment over that foray pulled Roger back in again, and the subject was dropped.

By the end of June I was talking with two or three of Bruce Weintraub's friends, but of course not calling him directly, since he still wanted privacy. I felt I had to get through to Bruce what I knew of the antivirals, especially when I heard he was doing a lot of investigative calling on his own. For a week or so it became a kind of circus, with Bruce fielding me questions through an intermediary. Finally, on the day of Dose 6, as we were leaving for dinner at the Perloffs', Bruce called and said, "Paulie, I know you know. Now can we talk about suramin?"

Nobody else has ever called me Paulie. Though Bruce and I had never been very close, running as we did in different circles, we used to joke at the gym two or three times a week—jock buddies. During the *Scarface* shoot, when Bruce was working on the art direction and I was writing the novelization, he'd feed me descriptions of all the Miami baroque excesses and then tell friends he was decorating my novel. He loved Hollywood to distraction and crowed with pleasure from the heights of his own career. He always seemed possessed with the speeded-up sense of time that attends a film in production.

Whatever it was between us, we connected on the treatment issue like nobody else I dealt with. Bruce managed to get through to the head of the CDC, the director of NIH, the doctor who wrote the suramin protocol. He had tried to break into the suramin study at UCLA with every connection he could muster, though he hadn't got the ticket yet. The problem was that he'd had both PCP and KS, and none of the protocols would bend that far. Bruce used to toss off a haunting image to underscore his desperation. "I'm a rotting pear," he'd say. "I have to have it now."

Of course I couldn't tell him Roger was on the drug, but we'd talk for hours about this antiviral and that, the side effects and what the next generation of drugs would look like. In the course of our daily bulletins, Bruce was the first person I ever heard mention AL-721, the immune-boosting agent at the Weizmann Institute in Israel. And later he fired the first shot in the battle for AZT, knowing about it before any of the doctors at UCLA, before the AIDS underground even. Because of Bruce, Roger and I started fighting for AZT early enough to get it. In that sense I owe Bruce the last ten months of Roger's life.

In the end he would have said it was only returning favor for favor, since Bruce gave me the credit for getting him on suramin, when all I did was put in a word to the right person at the right time. But it shows how deep the bond can grow among those of us on the front lines. Meanwhile I was able to drain off whole swamps of anxiety in my long, rambling afternoon talks with Bruce. Perhaps it simply had to do with being a local call, for Bruce and I could tie up the phones for an hour, talking between our two hills, whereas with Craig and Cesar the meter was always running. But I think it had more to do with Bruce's anger—his black, black humor as he excoriated the government, the drug companies, his friends who were turning away from him. We foamed with rage, the two of us. It was a neat irony that the picture of Bruce's released that summer was *Prizzi's Honor*, an entertainment perfectly suited to the dark idiocy of death that had caught us all up.

I even began to work at something, queer and tentative though it was, and not for public consumption. I'd been pissing and moaning for months now that nothing I'd written would survive. It was all out of print, and altogether it seemed a mere curiosity of the larky years between Stonewall and the plague. Since I couldn't seem to read anymore either, the irrelevance of books would sometimes sweep over me like nausea. How it began I don't recall, but I grew more and more fixed on a memory from Greece: those broken slabs and columns lying in the fields, covered with Greek characters erasing in the weather. I remember turning to Roger on the high ledge of ancient Thera, pointing to a white slab tilted in the earth, the lines of Greek barely visible now, and saying: "I hope somebody's recorded all this." And suddenly realizing the fading block of marble *was* the recording. How would committing the words to paper or floppy disk keep them longer than marble? Soon I was brooding that nobody left written artifacts anymore to slab the fields of the future. Out of some disjointed longing for ruins, I decided to make an artifact of my own.

I scoured secondhand furniture stores till I found a nice low table with sturdy legs. Then I went to Koontz to buy some paint, and only there decided it had to be blue. Aegean blue, I called it, remembering the window frames and shutters of the white stucco houses of the islands. I bought the most indelible felt-tip pen I could find. Then I set myself up in the garage and painted my Aegean table two coats of blue. When it was dry I began to write all over the surface, neat block letters stark as Greek, even on the bottom and up and down the legs.

Though the subject of the table appeared to be that poetry left nothing in stone, the content was scathing and rather antic. I spent three days on it, loving every stroke of the physical labor. I ended up giving it to Marjorie Perloff, who uses it now as a stand for her copying machine. Right away I embarked on a second, for Susan Rankaitis, an Aegean bedside table, with a pull drawer and a scallop detail. I envisioned a whole motel room full of blue furniture, with a continuous poem running over every square inch, no beginning or end.

On the Fourth of July we packed a hasty picnic and headed out to the Palisades in Santa Monica, arriving just as the fog burned off.

We spread out on the lawn under the palm trees, jogger madness all around us. In the pictures I took there, Roger looks a bit puffy and rather peaked, but the puffiness must have been genuine pounds, since he wasn't on any medication that would've retained water. On the way home we stopped off to see the Perloffs, and Joe took pictures of us laughing. Seeing the two of us side by side, I can tell right away that Roger's laughter is genuine and full, while mine is a kind of mimicry, as if I don't quite get the joke. It makes me want to turn and wrap my arms around him, so I can feel the quick of the real thing even as it wells from him.

Our last stop was Bel-Air, where we had a brief visit with Sheldon. He'd always been difficult to pin down for a visit—after half a dozen attempts to set something up by phone, one usually gave up for months. That was the rhythm of not seeing Sheldon for as long as we'd lived in California. We tended to glimpse him as most people did, at public events. Yet we would sometimes catch the wind right and manage to get him alone, and he'd ramble for a couple of hours, sharp and funny. Roger loved such occasions with him, but July 4 wasn't one of them.

Aggressively Sheldon kept asking questions about our careers, about which I for one wasn't interested in the least. I probably would've talked about my blue tables, I may even have tried to, but money was the only career Sheldon fully countenanced. For a while he simply deflected my AIDS questions, but at one point I brought up Bruce. He turned on me and hissed: "I don't want to talk about it, don't you understand? I've had enough of it!" I can't express how icy cold I went inside. This was the asshole who'd fenced us into our secret, and now he wouldn't share it with us? It's a battle scar I can still feel, and we saw almost nothing of him for the rest of the summer.

That week we heard about Barry Lowen, a TV executive and art collector whom I knew only casually. A year and a half before, I'd sat next to him on a plane, talking his ear off about vintage photographs. He had the sort of vivid, cocky good health of a tennis player in his late forties, not an ounce of fat on him. You hear about someone like that being sick, and your mind starts to waver, like the sketch they show you in Psych 101 that can be seen as either a young girl or a hag. There is the image remembered from life, a man in his prime just off the tennis court. Then something in your vision shifts, and you see the Other: housebound with the shades drawn, emaciated, breathing hard. Out of gossip and your own fear, you imagine the terrible changes. Roger and I were browsing one night in a bookstore, when an actors' manager we knew stopped to chat. He bragged about being one of the small circle Barry Lowen was willing to see in person. "The way he looks now, he's very selective," the manager remarked with a self-satisfied air. Always another rung of cachet in Hollywood, and always another arbitration for credit.

At Dose 7, a new man entered the study—Rick Honeycutt, a psychologist in his mid-thirties with a classic surfer's grin and the energy to match. He looked better even than Appleton, and considered himself in luck that he'd made it into the program so early. He had been with a lover for seventeen years, but he freely admitted he'd done a lot of playing in his time. Thus he figured he might as well be philosophical about the consequences. Those who are still early talk a different game from those who are not so early, and philosophical is a state of flux. I have a friend who's seropositive, on AZT and stable so far. He said to me last week, "You and I are close now because we're both in the same category. If I get really sick I'll just be bitter that you're still well." Too cynical, I thought, too lonely and bleak—but I know what he means. It was that sort of feeling that made me want to tell Honeycutt to shove his philosophy.

Though I have to say, we were a cheery little band now in the room that looked out on the banyan tree. Appleton at eight, Roger at nine, Honeycutt at ten, with enough of an overlap for us all to compare notes. The only consistent reaction to the drug seemed to be fatigue and a slight fever over the weekend. The nutritionist had come to know us all so well that she brought me lunch too.

On July 8 I called TWA and checked on flights to Chicago and Boston for the beginning of August. The trip would have to be engineered to the decimal point, so we could leave after a Friday dose and be back before the next one. My parents had been asking me to come home to Massachusetts for several months now, and I'd pretty much run out of excuses. Roger wanted to see his niece and nephew, and Sam had been urging that we get away, if only to prove we could. We ran the idea by the doctors, and they saw no reason we shouldn't go. Still, it was a very big step to take, booking those reservations, freer than we'd dared to be since the verdict. I think I was too excited about the prospect, too awestruck by the logistics, to waste any time wondering if this was the farewell tour.

On July 11 the Ferrari doctor called to break the bad news. My blood-test results showed that my T-cell ratio was reversed in the classic fashion, indicating exposure to the AIDS virus. I had not elected to have the antibody test specifically, because the sense of the community was so strong about the civil rights issues that might ensue. What if somebody started keeping a list? Lists were the first step to protective isolation, the polite term for camps. And what if insurance companies started red-lining those with the virus? If your numbers are in bad enough shape, you don't really need the antibody test. Or to put it another way, Ferrari no longer had a lot of bullshit to explain away my swollen lymph nodes.

My ratio of helper to suppressor cells was .5, where normal ought to be 1.0 or higher. I had 590 helper cells per cubic milliliter of blood, and though this was considered to be in the low normal range, it was nothing to write home about either; 1,000 and above was normal. Instantly I called Roger at the office to tell him, and he calmed me down and said he'd run the numbers by Peter Wolfe. We never really knew what Roger's own T numbers were, though at one point we were given to understand his T-4 was under 50. Best not to press the point, we decided. We had enough angles on Roger's bad news already.

I went ahead and had lunch that day with Carol Muske at Bistango, and though I didn't blurt my numbers, she recalls how wired I was. Carol was in the same position as Dell Steadman and several others, suspecting Roger had AIDS but talking around it, letting me say what I needed about the awfulness of things in the abstract. Carol had just put her third book of poems to bed, while I hadn't written a line of verse in ten years. We got to talking about who the audience was out there, and I told her about the tables, addressed to no one at all. Carol was suspended in that fugue state after a major push of work, numb and deflated, such that her own doubts about how to go forward happened to dovetail with mine.

We wondered if it was possible to write a poem that never thought about being published at all, or about reaching an audience. But then who would you be speaking to, just yourself? I don't know which of us first proposed the idea, but I know the phrase was Carol's. What about a "conspiracy poem" that would pass back and forth like a secret between two voices? We played with the notion, at one point considering telegrams, at another using a code. It was purely a lark, no pressure at all, just saying we might toss a few lines back and forth. Nobody else need ever see them. Though we left the project so wide open it could have ended right there, it made me think for the first time about putting a toe in the ice water of the imagination again.

When I got home Roger had already talked with Peter Wolfe, and he reassured me that I was in no danger. Peter had cited a study showing that 98 percent of those with a T-4 count of over 300 had not progressed to full-blown symptoms in eighteen months of tracking. This equation would prove to be very fluid indeed as the months sailed by, a kind of litmus test of *not yet*. For a while doctors considered an opportunistic infection to be imminent if the count fell below 200, but some people proved tenacious, and the low-range theory fell apart. Meanwhile there were men breaking through with KS at 600 or 700. The numbers as always are gibberish. What I remember most about that day was the rock-hard certainty of Roger's voice, and the flood of love I felt at his loyalty and concern, assuming control the minute I started to flail. In that glimpse of him I think I see something of how he must have looked at me over the next fifteen months, whenever I stepped in to fight for him.

Saturday, Dose 8. Roger having slept away most of Friday, we took a drive up through Beverly Hills to Franklin Canyon, where the tract châteaux leave off and the chaparral begins in earnest. We were trying to find the reservoir, an amoeba of blue on the map. It was a fiercely hot day, and all the scrub hills were straw and sage, no rain since April. The reservoir proved elusive, fenced off so effectively as to prevent even a flash of water. What we found instead was the tiny headquarters of the Santa Monica Mountains National Park, a bungalow with a ranger lady, grass snakes in glass cases and crayoned Smokey the Bear posters. We were the only visitors that afternoon, drifting among these mud-plain exhibits, absorbed as if we were combing the Athens Museum.

Later we went outside and sat in the shade of a sycamore. There we both had a rush of the throat-tightening sweetness of things, the perfection of time in which nothing at all was happening except that we were together. Wouldn't it be grand, said Rog, if it could all just stay like this? Sleeping it off dose by dose, gathering back to life and willingly giving up any claim on the Nile or the Left Bank, only to sit and listen to the dry leaves of a sycamore clattering in the breeze.

I had turned in *The Manicurist* over the Fourth of July, and there was a brief flurry from one of the producers about how perfect it was, especially its plot points. I had delivered in spades on page 26 and page 90, so for the moment the whole enterprise was judged to be hilarious. When you go unproduced long enough, you know that this is the only time a script is still warm, before the studio starts the autopsy and heart transplant. I remember when it left my desk the desk itself seemed to heave a sigh. My work had already taken a sharp turn, though I was hard put to categorize the medium. I'd never worked in the garage before, or back and forth in the mail.

Conspiracy: literally, breathing together. Every week or two now, a poem would go in the mail to Carol, and one from her to me. We began in completely different places, shooting in the dark, but quickly felt our way to a workable form of address, a courtly sort of confessional. My conspiracy lines—like my table thoughts—were all about the calamity, though for a while I couched my terms. I wrote about the white-stripe snake in Franklin Canyon and his "one medium mouse a month." About being allergic to bees, and the cloud of killer swarms advancing toward Texas. Shot through every fragment are undigested references to those who died young.

*Van Gogh was 36, Poe 40*
*Also Champagne Scott*
*and Frank O'Hara one of ours*
*no art news there*

I was writing with a very blunt instrument, but groping at last toward leaving a record—"to say we have been here."

It wasn't exactly a conscious choice to write about AIDS, yet the privacy of the bargain with Carol gave me the freedom to close in on it. And there was an unexpected return: Gradually there began to reassert itself that delicious balance between Roger and me, home for the evening after work. Looking back now, I realize I am selectively shrugging off the countless moron meetings Alfred and I were having at the networks, being turned down in thirty-one flavors. But I was inured to all of that now. If it used to require learning not to flinch when someone spat in your face, I didn't even feel the spray anymore. The bracing thing about this new work was Roger's enthusiastic response. He'd been asking me to go back to writing poems for a decade. He put no value judgment on anything I was doing—never did—he only tried to convince me it *was* work, tables and secrets and all. It didn't matter to him if I got another studio deal, for he was doing the nine-to-five work every day. I think that must have made him feel terrific, like the old days in Boston when he'd walk across the Common to Herrick & Smith, while I sat home shadowboxing a novel.

On July 18 I had lunch with Susan Rankaitis, whose abundant cheer was glowering now as she worked through a transition between one body of work and another. She'd had a major show at the County Museum a couple of seasons before, huge photographic pieces layered with shard images of jet aircraft. She was openly in a state of turmoil, questioning all previous ideas, and wanted to hear about the dead end I'd been in. Haltingly I explained that I needed to find a voice to witness the nightmare, trying to tell her all the truth I could but keeping it focused on Cesar and his two years at war. Then I talked about Marjorie's table and the collaboration with Carol, and over coffee we began to speak of a collaboration of our own. Susan had been wanting to work with a human figure and explore the image of Icarus falling—the earliest manned flight in the books. I offered to do some lines on the same theme, then found myself agreeing to pose for the piece.

Five days later Roger and I went over to Susan and Robbert's for dinner. Susan had been mugged that afternoon as she was getting out of the car with the peach pie she'd bought for dessert. Though her window was smashed and her purse snatched, the pie escaped unscathed, and we polished it off. Then we repaired to the studio, and I posed naked as the plummeting boy. A man of forty running to flab, his youth beached like a whale, makes a very unconvincing boy, but Susan assured me that I would do fine. She works by manipulation of photographic imagery, exploding things for their shrapnel value, the surface effect metallic, positive and negative at once. Mostly she wanted the right silhouette pose as the winged boy feels the burning sun and his wings fall away.

Roger and I would watch the piece grow monumental over the next several months, but I recall the evening of the photo session as the opposite of the night two years before with Jack Shear. I could feel the lumpish dislocation of my body, ticking away as I did these

Isadora Duncan tableaux, my arms up in front of my face. I had broken through to Diane Arbus status, and I didn't care. Meanwhile Roger worked quietly as Susan's assistant, adjusting lights and holding equipment, placid as he had been at the previous shoot.

At the end of July a couple of friends were in town from Philadelphia. Joe and Stuart were both comfortably ensconced in chairs of English, and they'd been together forever. They reported that a mutual friend—Ed Tompkins, a Washington lawyer—had been unable to join them on the trip to L.A. because he was down with a virus, something to do with his nervous system. My face went blank as I stared at the diagnosis, but neither of them appeared to recognize the naked truth. Apparently there had been tests that proved nothing conclusive, but Ed was probably covering up. He was a shy and closeted man, with only the most tentative experience and a single love gone bad. He'd had the dubious distinction of being pursued by a member of the White House staff, that closet within a closet, but Ed had turned him down because the man was married. When Ed died three months later, Joe and Stuart finally admitted it was AIDS, but still in the face of the ghastly denial of Ed's family, who kept the gay friends away and let no calls through as he lay dying.

By Dose 10 we had put in the good word for Bruce, and he'd been accepted into the program. Roger and I simply asked if he could be scheduled for some other day than Friday. Bruce was so eager to start he was beside himself, and he kept up a flow of good news from all his various sources. Bruce was really Suramin Central, much more than I. His sister Carol came out from New York to visit, and we had the two of them over for Saturday lunch. It was the first time I'd seen Bruce in four months, and he looked okay, if a little thin. We mostly talked about other things. Carol told me months later how thrown she was that weekend; that lunch with us had been a kind of haven from the nightmare, proving that she and Bruce could still laugh. After lunch we sat in the garden, and Bruce waded in at the shallow end of the pool. He had been my gym buddy for years, strong and street tough and speeding with energy. Thus there was something terribly poignant in seeing him balk at a swim, saying he was feeling a chill and wouldn't go in any further. He seemed suddenly modest in his body, he who always crowed and darted about. It was all going to be fine, though, once he started on suramin.

On the Tuesday after Dose 10 Roger was running a fever, and Cope ordered a blood panel and blood-gas test. Just going upstairs again to the pulmonary unit was terror enough, and I started to go out of my mind again. But the tests proved negative, the fever disappeared, and we wanted so much to believe it wasn't the suramin causing a problem that we blocked the thought. It couldn't be the suramin, because then there would be no magic bullet. Now, in the last week of July, a wave of AIDS stories seemed to cluster on the news, all of them bad. We heard that Rock Hudson, flown from Paris on a rented 747 and brought to UCLA by helicopter, had only a couple of months to live. It was announced that dentists must start wearing masks. The first whine of panic began to mosquito the airwaves. There was never a word about antivirals or any other treatment. The "always fatal" illness, they said.

But at some level it couldn't get us down anymore, not after nearly three months of healing breezes through the banyan tree. On Friday, August 3, Roger came home from work and discovered he'd lost his watch, which had always seemed as grafted to him as the sapphire ring. He called the CRC, the office, the restaurant where we'd had lunch—no luck. As we nestled in bed that evening he let the watch go, with a mournful observation to the effect that things after all were nothing. Then we got up next morning to drive to Laguna, and the watch was in the front hall on top of his briefcase. The pin in the strap had worked loose. "Oh, the curse sometimes lifts!" I hooted in my journal, embracing all evidence of false alarms.

I worked double time on Susan's table during the days before we left for Chicago and Boston, the subject of its text being the art market and Neo Ex. By the time I was called in by the studio for the evisceration of *The Manicurist*, I was already very far away. It was a particularly savage and ugly meeting, with Whoopi's people screaming that she'd become a secondary character to the guy, and the studio executives demurely let me take the heat as if it were all my bad idea. This movie has no hook, I was told. It had no setup, it wasn't character based and it wasn't funny. No one had a kind word to say about page 26. Yet the hand that jotted down their pointless notes was flecked with blue from the table painting and belonged to a man who didn't care anymore.

We hadn't had a suitcase out of the closet since Greece, and the sanity of packing for two was terrific. We'd always worn the same size shirts and underwear and socks, so nothing belonged to anyone in our house. We were always disguised as each other anyway. I raced over to Susan's the day before we flew so I could present the table. I always like to leave matters as finished as possible before a flight, though AIDS is a remarkable cure for fear of flying. We were definitely not going to die in a plane crash. This is another way of saying something Star once wrote me during her seven years in Asia: *The cure for metaphysical pain is physical pain.*

The schedule was tight. We were due to leave for Chicago on a noon flight, and I didn't sleep with excitement. We were early for Dose 12 and talked about our trip with the CRC staff as if we had a week's pass from the battlefront. We raced home to meet the car that was taking us to the airport, barely five minutes to spare. As we came up Kings Road I spied our mail carrier in her Jeep, and I braked on the hill, got out and gave her a bunch of letters.

When I tried to start the car again, it had frozen in gear and wouldn't move. So we had to leave the Jaguar sitting out on the street as I ran up to the house, called Jaguar service, grabbed the bags and bolted to hail the airport driver. We always traveled madly, one well-laid plan going haywire after the next, but that after all is how a trip turns into a journey. All we knew was that we weren't hostages anymore. We even had to laugh as we passed the mute Jag, dead on the hill, because there it was again in case we had missed it the first time: Things were nothing.

# ·VII·

It was a long flight, with a stopover in St. Louis, and the crowded plane and terminal chaos were daunting. Sometimes you just had to throw up your hands. There was no way to protect yourself from the germs of the teeming summer masses, except to touch the surface of life as little as possible. Al and Bernice picked us up at O'Hare, and we stopped at a deli in Skokie on the way home. When they marveled at how well Roger was looking, both of us could feel them relax at last about his recovery from the pneumonia of the spring. Roger and I slept on the Hide-a-Bed in the den, on an inch-thin mattress that felt like overnight camp. I remembered visiting the parents in '75 on the way to California, when they didn't understand we were lovers and put us in separate rooms. Now I was family.

We were hustled up early and over to Jaimee and Michael's where the children roared with excitement to see their uncle. Six-year-old Andrew trounced me at tetherball, a game I leaped to play only because I didn't want Roger straining himself. Michael had arranged to borrow a friend's boat for a cruise on Lake Michigan. All eight of us piled into Al's Cadillac and drove to a marina in the city. Four-year-old Lisa sat on Roger's lap, and they laughed and chatted happily—while I despaired of keeping the children from breathing in Roger's face. But even I managed to unwind as the boat got under way, a thirty-foot Chris-Craft that looked like a rocket and slept six.

We headed up the sapphire lake, the sunny day cool and dry, none of the choked humidity we expected in Chicago. As we passed through a choppy wake on the way out of the marina, Al suddenly gripped the rails and ordered Michael to turn back. He was scared of deep water because he'd had a brother who drowned, sixty years ago. In the peculiar way that families accommodate their unreason, they all benignly ignored Al's expostulation, and after a moment the phobic spell passed. Ten minutes later he was serene as an old salt, beaming at his two generations of children.

Roger had a glorious time. I have a great picture of him grinning at the camera in his rumpled sailor hat, and there are no qualifications in the evidence. He looks completely well again. Though sleepiness and fever had been dogging him during the Saturdays after the suramin, he was unaffected today, impish and laughing. No wonder they all thought he was out of the woods. He sat by Jaimee, both with their arms folded against the buffeting of the wind, talking close to be heard above the roar of the wake. They looked like brother and sister, no other way to say it—brother slightly older, sister all ears. But perhaps the memory is so vivid because Andrew and Lisa were a microcosm of the same. On the way back the children grew restless, being confined so long in a small place. The two of them bounced around the car as we headed back to Glencoe. Parents and grandparents both were accustomed to the ruckus, but Rog and I wilted like maiden aunts and couldn't wait for a nap.

There were no very sad times in Chicago, though Sunday morning as we dressed for breakfast Roger shook his head and said in simple wonder: "I don't feel young anymore." I of course started to cry, which is rough on an empty stomach. He didn't look old or seem old then. I suppose some of it had to do with visiting his folks, where a man is always slightly out of tune with a vanished child playing in the next room. What made the moment especially hard was to think how boyish Roger had always been. Half the time we were like twelve-year-olds, and the world out there was a sort of field trip. He had the energy and sense of mischief of a seventh grader with all A's, though he also had a sweet tooth for playing hooky. Since I had been such an ancient child myself, gloomy and bookish, the only kid I ever got to be was with him. So when he said his youth was over, two children seemed to disappear into the woods, hand in hand like Hansel and Gretel.

Sunday afternoon we drove into the city with Al and Bernice and took a long walk around the downtown area. Al, who is no fan of L.A., busts with pride about his big-shouldered city. We passed the spot where H & H Restaurant had been, and they laughed about the time Roger managed it for a couple of weeks while his parents were in Europe. Then Al steered us on a tour of the city's sprawl of outdoor sculpture, Nevelson to Dubuffet, showing each piece off as if he were part owner.

During dinner at Jaimee's, Michael took her aside and said, "Why is Paul like a mother hen around Roger?" And later, to me: "There's nothing wrong, is there?"

That night I asked to see the pictures of Roger's bar mitzvah, and Bernice gladly pulled the albums out on the dining room table. I'd never seen him that young before. He had the most unabashed grin in his thirteenth year, and Jaimee in all the flounces of a party dress couldn't hide the tomboy. As we leafed through the mid-fifties, all its catered optimism, there were groups of people scarcely older than Roger and I, arms around each other and grinning at the camera. "He's gone," Bernice said matter-of-factly, pointing at this one and that one, "she's gone, he's gone. . . ."

Monday morning we were off to the airport, proud of the visit, the family content, the Chicago branch at least. Halfway to Boston, I started to cry—

> *that I had nothing happy to report, that I was only going to have one bad ugly talk after another, that I was so sad and so finished.*

But things were easy with my parents right away, due to the fact that there was so much mutual affection between them and Roger. Since we moved to California I had taken to visiting them alone, and without Roger the days in Andover always made me feel as if I'd never escaped. I don't have quite the grown-up relations with my parents that Roger enjoyed with Al and Bernice. I'll take a lot of the heat for that, since I'm not very forgiving about the wrongheaded notions my parents once had about gay, or their anguish at having produced a writer. ("It's fine that you want to write, but what are you going to *do*?")

Nevertheless, they are very decent and giving people, plain Yankee folk whose fences make good neighbors. The fly in the ointment for me had always been my mother's Christian fervor, a long-standing matter of locked horns between us. Low Episcopal, we're talking, not snakes and tongues and Tammy Faye. Christ Church in Andover was a fount of liberal outreach, shining with irreproachable convictions, yet my mother's sprinkling of God in every conversation had created a cloying atmosphere from which I kept my distance. All the same, the longer Roger and I were together, the more we healed as a family. It's not an accident, I think, that neither of us came out to our families until we found each other. Alone it is hard to want to face the barrage of clichés, and the closet is so much easier. But you can't go on very long hearing your heart's deepest core called your roommate.

After dinner the first night, Roger went up to rest, and I had a pretty good talk with the parents about AIDS. They were reasonably well informed, though they clearly didn't think it could touch us. Then I broke the news about Cesar, and I could see their assurances falter.

Tuesday I took Roger to Phillips Academy, where I endured the existential acne of high school and later taught for five lambent summers. I took two pictures of Roger on the great lawn in front of the art gallery, the two I keep closest to me now. In both shots his arms are open in a great embrace, and he's laughing with pleasure, the humid green of summer in the elm alley behind him, the sky milky and palpable. We walked from there into the Cochran Bird Sanctuary, a walled enclave like a private forest, where I'd taken a thousand solitary walks as a youth, winter and summer. We made our way to the pond and sat on the stone bridge above a lazy brook fanning its algae. We talked calmly about all we had and how we were doing fine. Then Roger looked up into the trees, and a choke came into his voice: "But what if I die?"

"You're not dead," I retorted passionately, not quite addressing the question, though at the time it seemed the only answer. "We're *here*. We're going to win."

I believed it absolutely then, that we would lead the way. Soon the antiviral news would break, and the hope would come flooding in. We would be there to show the rest how to bear the joy. Now of course I can answer Roger's question in endless sad Keatsian detail, but at that heightened moment I hardly took it in. Mid-August in the sanctuary was the peak of summer, gaudy with life. No wonder the grasshopper laughs at the ant. I've been on the stone bridge only one time since then, about three weeks after Roger died: snow and cold, a sky that smoked like dry ice, and no birds sang.

My parents had arranged to take us to York Harbor in Maine for dinner, a favorite place of theirs on a spit of land at the mouth of the York River. When we arrived I settled Mother and Dad in the restaurant, then Roger and I went out to the beach for a breath of sunset. We walked down into a sort of cove with a rim of summer millionaires, white shingle with green and blue shutters, the opposite of Aegean. Roger stood, feet apart, in the sand, sniffing the sea breeze while I capered down to trail a hand in the water.

When I came back with a smooth gray stone the size of a silver dollar, he was serene with delight. We talked about Proust and his grandmother, the seascape frieze at Balbec. I'd always hated Maine—too cold, too WASP, you blink and the summer's gone—but I had to admit the northern light was exquisite today. Yet I said I could only enjoy it because he was there with me. I didn't particularly mean because he was still alive, rather that I wouldn't have enjoyed it half so much if I'd been with my parents alone. Or worse, all by myself: I'd stood on enough solitary bluffs to last three lifetimes. But Roger's eyes welled with tears at my words, and I wanted to scream with stupidity, because I hadn't meant to make him sad. Yet sunset is so mercurial, it changes in front of your eyes. I remember us leaving the beach laughing, shoulder nudging shoulder as we furthered the larger conspiracy.

Wednesday Roger went into Boston to visit old friends. I had lunch with my aunt Grace, who'd lost her husband about five years before. She said she missed him now more than ever. All afternoon I kept calling Rog in Boston, suddenly feeling trapped again in a small town, neuter as an old schoolteacher. Roger was very emotional. He'd had lunch with Miriam Goodman, a woman he'd known since Brandeis and the poet preceding me in his life, but he didn't tell her. Then he went to Tony Smith's, on Brimmer Street around the corner from our old place on Beacon Hill. Tony taught political science at Tufts and had been one of Roger's best buddies in grad school, the only one who was gay. Tony asked about the pneumonia, but not really very concerned, and Roger tried to be stoic and dismiss it. Something in his voice made Tony turn from the stove, where he was cooking: "But you're all right, aren't you?" Roger shook his head and started to cry.

I wanted to go in then and be with them for the evening; it was only a half hour away. But my mother was having a bad asthma attack after dinner, so I stayed home and waited. I played cribbage with my father to make the time go by, one eye on the clock. Then out of nowhere Alfred called from L.A., telling me CBS wanted to make a deal on a story of ours. My parents were elated, and I mimicked their excitement but felt myself hoarding the good news for Rog. I was only excited about telling him, not about the thing itself. I loved how thrilled he was for me when he got home.

After midnight the two of us sat out alone on the back porch in the ink-green quiet, and I couldn't stop crying. I was just so glad to be with him again; the half-day alone had left me a wreck. Yet the whole trip seemed to release a bottled-up flood of tears, and a thousand things hit me sad. At one point my mother was reminiscing about what a darling baby I was, happy and verbal in the extreme, and all I could think was I couldn't think back that far. I always got stuck around eight or nine, my brother in Springfield with casts on both legs, surrounded by bewildered toys. And then she told me something my grandmother used to say: When the children are safe in bed at night, those are your best years. Packing to leave on Thursday, in the bedroom under the eaves where I'd slept off and on for thirty years, I blubbered to Roger, "I'm crying because our parents can't get us out of this."

The flight home was interminable, rerouted to Canada because of thunderstorms, which meant we ran out of fuel and had to pit-stop in Vegas. During two hours on the tarmac with no air-conditioning, I could practically smell the germs as the sealed air thickened and choked. We were due in at seven and got back at eleven, but made it to the CRC next morning for Dose 13. And heard that Appleton wasn't feeling well that day, so it was decided he wouldn't come in. To miss a dose of the precious liquor was simply unheard of. It seemed immoral to let that bed go empty, but if you're racked with diarrhea what are you going to do? It remained unspoken whether or not the suramin could be causing it. And week by week, as Appleton stayed out sick, I grew more and more pissed off that he was screwing up the curve.

Against all that there was Bruce's optimism, as he would tell in exquisite detail about his appointments at the CRC, not realizing I knew the place far better than he. At this point they hired on a nurse-practitioner, Suzette Chaffey, to oversee the study. There was endless gossip among us moonfolk about this perky lady. She had spent time abroad as a child and worked in Switzerland, as I recall, so she had about her a European savvy that Roger and I found irresistible. Besides, she was somebody new to tumble out all our Ancient Mariner tales of fevers and flagged hours. Since she would ultimately come to oversee AZT as well, we were to travel a long road with Suzette.

Joel called in mid-August from Santa Fe to say that Leo was doing terrifically well. They were living high in a piñon forest above the city and had become vegetarians. Leo had had the good fortune to be accepted into the suramin study at another hospital in L.A., so he would be flying up every week for a dose. It was clear that Joel would not be accompanying him. In the most offhanded way, during this Pollyanna blizzard of good cheer, Joel happened to say he was negative—meaning he didn't have the virus.

Here it gets very subtle, because the fiction is spun as fine as Scheherazade. "You know, Leo and I never fucked each other," he said. "In fact we never had much sex at all." I thought: He's getting ready to split and wants to look clean for the next one. A man on the make has to set the record straight.

There is such infinite variety about the way people tell you their negative status. Some are openly full of remorse and feel they have failed you. A man I've never met wrote me after an interview I gave about my antibody status:

*When my test came back negative last month, I was overwhelmed with a sadness I hadn't expected. Coming back alive is a guilt, a terrible betrayal, a necessary starting point.*

Some hoot with excitement and forget you might not be as thrilled as they. I have friends who will not be tested at all because they know how shamefully glad they'll be. Their gut instinct is they're negative, so who are they trying to kid? They are always the first to tell you to stop being so AIDS-related. *Lighten up*, they say.

My own consistent opinion is selfish enough and sounds suspiciously Orange County. I want the two million—or the five million, depending on whose scenario piques your fancy—to have themselves tested and know, so I will have people to talk to. Because after midnight and during weekends I cannot talk to those who play at business as usual. I want to tap into the rage of the positives so we can throw buckets of sheep's blood on the White House lawn and spit in the faces of cops with yellow gloves.

How tired was Roger on Fridays? It's very difficult to assess, because that was the focus of our denial now. Appleton's weeks of diarrhea were something apart and ARC-related. The continuing cautions of Craig's researcher friend in Houston didn't somehow translate to the safe haven that looked out on the banyan tree. The sample was so minimal, after all, with only a hundred or so on suramin protocols throughout the country. Peter Wolfe was too busy now to be spending his mornings with the Friday club, and Suzette didn't seem to be privy to the latest data. Mostly we tried not to worry about the drug, because we had enough to worry about the disease.

The major drug reaction was rage anyway. It made us crazy to think the FDA or the NIH hadn't made funds available so that thousands could be on the drug. There was desultory talk in Washington of an HPA-23 study, when we all knew in the AIDS underground that the drug was useless. How could suramin still be the only game in town? We understood already, even as laymen, that what was needed was a new antiviral, especially one that would cross the dreaded blood-brain barrier. The gathering evidence during that summer indicated that a higher and higher percentage of AIDS patients were suffering from effects of the virus in the brain. The fear was that an antiviral that controlled HIV in the blood would somehow send it ravaging into the brain, where most drugs couldn't follow—including suramin.

By now I was fielding calls from all over, friends of friends saying they understood I knew someone on suramin. I would give the status report as hopefully as I could, sometimes referring them on to Bruce, the indefatigable spokesman. The burden of my own message was that everyone must start demanding these drugs, because the system wasn't out to cut the red tape. Indeed, red tape *was*—and largely still is—the system.

The rumors were appalling. It was said that everyone appointed by the Reagan administration in a major public health capacity was either a Mormon or a fundamentalist. The chief spokesman for the administration now was the overripe and venomous Patrick Buchanan, one of whose major qualifications for the job was his widely quoted remark that nature was finally exacting her price on homosexuals for having spilled their seed against her. The right-wing firebrands are obsessed with sodomy, always forgetting that half of the gay world is women. This deterministic smugness, whereby we were only getting what we deserved, was so widespread in the upper chambers of the government that the AIDS issue probably never darkened the threshold of the Oval Office. Not to mention the fundamentalists: Though the press would not report anything about the antivirals and wouldn't assess the scope of the death of a gay generation, they reported with loving detail every ranting speech of the Falwell-Schlaflys and their money-changing brethren. "God's punishment" was the major level of public debate in 1985: hate, it appeared, was the only public health tool available.

Toward the end of August, Cesar came down with Dennis for the weekend. I picked them up at the airport and immediately felt easier about the swimmer. He was witty and very self-assured, extraordinarily tender and deferential to Cesar, who was himself transported by the notion of putting together the alpha and omega of his life, love and friendship. He'd obviously been bragging to Dennis for months about his blood-brother ties to me, painting a sunstruck version of me that was hopelessly prewar. I thought I was pretty casual and breezy that afternoon, but Dennis started to question Cesar as soon as they were alone. Why isn't Paul happier to see you? he asked suspiciously. I knew the next morning that Cesar had finally told him the truth, because Dennis stared at Roger across the coffee with a stung look of desolation.

It was altogether a heartrending time, rife with the end of summer. Each night I would curl up beside Cesar in the back bedroom for the day's final gossip, as I had since I was twenty-five. Dennis was sleeping upstairs in the loft. But now the wrap-up moment at evening's end was skewed by the encroaching horror. The weeping from the open wound in his groin was so severe that the rank burnt-sugar smell seemed to pulse out into the room in waves. The lesions had spread between his fingers, and his body was peppered all over with them. Bravely he looked on the bright side: at least they weren't on his face.

And amazingly enough, he talked on and on about Dennis as if it were a growing and consolidating relationship. He was furious that Dennis continued to stay with the safe old roommate who used to be a lover but wasn't quite. "It's now or never," Cesar said. "I need a little commitment here!" It was so close to a delusion it took my breath away, but it wasn't quite. He was drunk with love, and the passion in his voice, the helpless spill of feeling, were on a wholly different plane from his body.

I told him to go slow and be glad of what it was—so easy is it to play with other people's time. On Saturday we were out by the pool, and Cesar swore he'd be swimming again by Thanksgiving. That evening the four of us went out to dinner, chic cold pizza at Bistango, the double date Cesar had been waiting a decade for. He was magic that night, the topspin of his stories dazzling. When we all came laughing home, Roger and Cesar both turned in, and Dennis and I sat outside and talked. When at last we came in, we hugged each other good night, and something about the moment turned the hug to stone. We clung together for what must have been ten minutes, saying absolutely nothing.

On the final night of their visit, some plainspoken friends of ours who'd always enjoyed Cesar begged to be able to see him. I said yes without giving it much thought. They were a long-hitched couple who'd just returned from a trip to North Africa, and knowing what a spellbinding traveler Cesar was, they wanted to trade stories with him. It took Cesar over an hour to dress his groin and get ready. As he painstakingly bandaged himself, something he did twice a day now, he looked over exasperated at Dennis and said, ever the linguist: "This is what it means in French, *quand même*."

We all went out for Indian food, and the mild folk told their African story, admittedly rather tame. I could see the flare in Cesar's nostrils as he asked, "But didn't anything unexpected happen?" My Uruguayan friend was simply not an angry man, yet I could feel his spine stiffen. "You don't still travel to go see the tourist sights, do you? What about the atmosphere?"

They tried the best they could to describe the atmosphere. Later, when they'd left, Cesar was scathing in his satire of them, no matter how decent they were, because they were well. Their bourgeois marriage, with its twice-yearly tours of the world's garden spots, mocked poor Cesar, who only wanted a little space for his own sweet season with Dennis. The other thing I recall from the Indian dinner was observing that I'd known Cesar for fifteen years, and him correcting me. "Seventeen," he said, reminding me of the first evening, talking Borges in a drawing room. No rounding off of years allowed.

When I took them to the plane next afternoon, so much still hadn't been decided, even whether Cesar would go back to school in September. Phrases like "leave of absence" were beginning to creep into his conversation, however. The other gay teacher at his school had been diagnosed with PCP that summer and had already dropped out. The rattled headmaster called Cesar's doctor and asked, point-blank, did he have AIDS, only to be told Cesar was being treated for cancer. Cesar still thought he could give it a try, but I suggested perhaps it was time to focus his full attention on his health. We got to the airport and had a last bear hug outside the terminal, as we had after two dozen holidays. "Thanksgiving," he promised resolutely. "Quince," I replied with a wag of a finger, and he and Dennis went off laughing. I never saw Cesar again.

"There's a kind of grace sometimes, isn't there?" Roger said when I got home. He meant the relationship with Dennis, knowing as well as I how fiercely Cesar had always longed for the eloquent partner.

I rack my brain to think what the signs were at the end of August, but nothing out of the ordinary. We were both working along, and if normal is too strong a word, then perhaps it's enough to say that compared to the nightmares we heard about, nothing much was happening to us. Roger's publicist client with meningitis died around that time, and Roger began the probate process. "He didn't have very much," Roger said simply, putting much more work into the estate than he'd ever be able to charge it. The mechanics of that probate, which I would hear about every now and then, evoked the specter of single men in one-bedroom apartments all over West Hollywood: the little collection of antiques, the closet full of labels and Hollywood memorabilia, the small life-insurance premium left to a sister in the Midwest.

"People do die of other things," a couple of friends have observed trenchantly when I've talked as if only the war still kills. Such a useful perspective. Of course I knew people were dying of everything out there, because now I read obituaries and never missed a beat of any other disease, from lupus to yaws. A woman I know who lost her two-year-old son last year says she never tires of reading books about other tragic children. This is called "being immersed," by those who are not immersed. Still, I'd reached a point where I couldn't shed tears anymore about people dying in their proper time. If someone had seventy years, I'd think: So what's the big deal?

Yet all it takes to soften such an attitude is someone you know. At the end of August the old Viennese doctor who lived across the street died suddenly of a heart attack. We were friendly neighbors to him and his wife, and I used to take the dog to visit Mrs. Knecht, who adored animals. A week after Dr. Knecht died, I hauled the dog over for a proper condolence call. I'd never confronted a widow in the full moment of loss before. Just then it seemed as sad as anything—even AIDS—that Rudolf Knecht, who'd escaped the Nazis by the skin of his teeth in '42, had died at eighty-five and now left behind this remarkable lonely woman. If by that time I had come to a point where I felt most comfortable talking with those in the war, it proved a natural thing to include Mrs. Knecht. I took to visiting her every week, and listened as she talked about what it all meant.

"Life," said Mrs. Knecht with a fine Viennese loftiness, "is like a curtain pulled avay from a vindow, and you see the beautiful landscape, and then the curtain drops."

Over Labor Day weekend a writer friend from New York who summered in Vermont wrote us a short and breezy note about all the odds and ends he was writing, reviews and essays and poems. He was not an unkind or insensitive man, and he would prove a caring friend, once he knew. But I felt a rage toward him akin to what Cesar felt for the two men back from Africa. Business as usual in the literary department was especially galling, and I'd already started snarling to people that no one should write about anything anymore but AIDS. I loved it when the Eastern Europeans complained at the annual PEN conference: Why don't Americans write about anything? Reagan would be remembered, I said, for just one thing, that he presided over the denial of the calamity. I'm not saying any of this was true or even coherent, only that it was what I thought as my rage came to consume my despair. And about my friend idly writing from the country as the roses faded in his Vermont garden, I said:

*those back east*
*with their head up their ass will all have*
*tenure soon.*

On September 5 there was a meeting to discuss the video presentation for the annual Center dinner. I was meant to put together a text for a slide tape that would show the wide net of the Center's reach in the community. The man who would handle the visuals was a supercilious type who noted at the previous meeting in June that he didn't know anyone with AIDS, only to be withered by a member of the board: "You will." Now we were trying to brainstorm a way to talk about the Center as even more precious and necessary during the nightmare of AIDS. We had to have a place to serve as a sanctuary for our gropings as a people, especially in a dark age when it would be harder and harder to be openly gay. For once we would not internalize the homophobia. We must remember and pass on what it was like when the community was effectively splintered by the closet, like dissidents in solitary.

It all sounded stirring enough, though I wasn't buying the reality. Brave men and women all over were starting the AIDS lifelines and speaking out, but my sense of the man on the street—Christopher, Montrose, Santa Monica—was of a growing loss of center. If you have enough barriers up, it doesn't matter whether the closet door is open or shut. As Rand Schrader drove me home from that meeting, I kept talking about the mushrooming numbers. "We're all going to get it," I said.

"Paul," he replied evenly, "there are only thirteen thousand cases."

Only. In one way of course I don't blame him, since I was such a broken record, and he was just trying to get his dinner program set. I know the L.A. gay community was responding to the caseload with money and passion. In part my estrangement was self-propelled. About three weeks before the meeting there had been a cocktail hour at a swank house in the hills just below Bruce's, with a pool that seemed to hover above the city. The two men hosting the party were bright and successful professionals, *GQ* profiles with a flash of Melrose funk. The purpose of the affair was very bald: The upper-income brothers were meant to announce their support for the annual dinner by volunteering to sponsor tables. In a scene that struck me as vaguely medieval, a hundred of us stood around the pool—agents, lawyers, doctors, realtors—and sounded off one by one.

I recall being pissed at everyone that night for being healthy and cheery and tan. I don't think Roger felt any of this. As the event went on and on, he sat down to rest on a chaise by the diving board, benignly calling out when it came his turn. I worried about how cold it was out there, and could he get a chill from the breeze off the pool. I actually liked a lot of these people; I even missed them in our lives. But we were on the moon, and they weren't, and we usually declined their invitations. I only brood about the moment now because the two men who gave the party have both been diagnosed since, and they've sold the house with the view to Catalina, and the *GQ* jobs are gone.

On the Saturday after Dose 16 we packed the dog and left for an overnight at Lake Arrowhead, in the San Bernardino Mountains. Despite all our forays into the California wilds, we'd never been to the mountain lakes, and that was reason enough to go. We were feeling cocky that day. The entry in my journal from Twin Peaks is the last of any length and with any spirit, the last in real time.

*We ended up in the dearest cabin in the woods—#8 Mile High Lodge, run by these two utterly improbable Bengalis—& though I locked the car in gear & nearly stranded us for days the fat man from AAA had us going in a trice & we headed for Blue Jay to a sane coffee shop for sandwiches & milkshakes. Puck's been a wreck, but we're glad to have him along & we've had good walks. The afternoon sun was glorious & the evening came down nice & cold. We walked along the lake and ate at Heidi's(!)—R didn't eat enough. I don't feel like a loser here. I feel escaped & alive. We passed 3 young girls in the parking lot at Heidi's & realized we were middle-aged to them, & we didn't give a fuck, not one fuck.*

Roger was pretty tired and napped a lot, but we paced ourselves for those mile-high walks and on Sunday sat out on boulders and watched the hawks. It amazes me now how whole life was, even at the brink.

That week I heard the final studio decision on *The Manicurist.* "Steve says he doesn't want to make this kind of movie," the producer told me coldly, acting as if it made his hands dirty just to be talking on the phone with me. "What does he mean by that?" I retorted, not quite sure what would happen to the rewrite money. "He means it's a piece of shit," came the reply, which in turn meant they had to pay me for a draft I never had to write. A boon of sorts, though they sent the check with the tacit understanding that they would break my knees before I'd ever be allowed on the lot again.

It was right after that Bruce called, full of excitement. He was six doses into suramin and holding steady. This was the bulletin: One of his myriad sources had told him about a new antiviral just beginning human trials at the University of North Carolina. Compound S, it was called, and there were two AIDS patients on it. Bruce had been unable to track down anyone else who knew about it. Next day I mentioned it to Peter Wolfe, who'd heard of the drug that very morning but knew nothing more. We drew a blank with several other doctors, and Craig's sources in New York hadn't heard so much as a rumor.

Then, embarrassment of riches, Bruce called the next day to report new data about the Israeli drug—AL-721, an immune-boosting agent that had been used successfully on a child with an autoimmune dysfunction. The Israeli researcher had told Bruce that the FDA was throwing up roadblocks to prevent them from testing it in this country. So now we had our new underground agenda, and between us Bruce and I made hundreds of calls to find out more, though still we had no major sense of danger about suramin. We were just trying to keep ahead and be in the right place for the next phase.

On September 10, Craig arrived from New York for a week's visit, primarily so he could go to Mexico to get a supply of ribavirin. It was the one antiviral that was available over the counter, though the counter was across the border. Craig was impressed and delighted to find Roger looking so much better than four months before. By now Craig and I were accustomed to the two-tiered policy of talking nonstop about AIDS when together but not around Roger. Next morning the two of us got up early and headed south, taking the Datsun rather than the Jaguar so as not to be conspicuous at the border, where we would be bringing over a thousand dollars' worth of drugs. In theory one was allowed to carry back enough for "personal use," but the area was very gray. There was talk in the underground of detention and confiscation. The mind reeled at the challenge of avoiding germs in a Tijuana jail.

"The Treasure of the Sierra Madre" was our nickname for the drug that day. So many have gone over now to get it that the ribavirin buy has become a kind of reflex. Everyone knows which pharmacies can be trusted, which are rip-offs. There is such an elaborate system of mules that I can usually obtain the drug these days with a single phone call. A friend who keeps a fair stock on hand meets me on the corner of Western and Santa Monica, outside Fedco, to make the swap. But in the fall of '85 there was still a quality of the unexpected about the smuggler's journey. It only reinforced our sense of being outlaws, and for once there was a tinge of romance to it.

We had the names of four pharmacies, and decided to go to one outside the city center, which involved a rattlesnake drive along the

Mexican side of the border. For a space of several miles we saw illegal immigrants pouring through holes in the chain-link fences, seven or eight in a family with trash bags full of their worldly goods. Taxis would screech to a stop at certain gaping holes, and the refugees would tumble out, wide-eyed at the port of entry as if Liberty herself had cut that fence.

The pharmacy was in a dusty town across from a bullring, in view of the green sluggish southern ocean, raw with the smell of kelp. We bought all the ribavirin and isoprinosine they had, chatting amiably with a couple from San Francisco who were buying cancer drugs. I realized then we weren't the only ones being driven underground by the FDA. We were part of the nether world of the sick, trying to get some control, taking risks the government wouldn't sanction, and all in the same boat.

We soaked up miles of atmosphere and were giddy as we waited in line at the border, trying to look proper and nonaddictive as we gestured toward a trunkful of drugs. We were waved on through and drove home into full gold sunset, exhilarated and in charge of fate. By the time we spun the story out to Rog it was already part of our history, something we had won.

Craig had asked if a friend could visit from San Francisco for the weekend. I agreed because I wanted to have some quiet time with Rog, figuring his friend would be a diversion for Craig, whose assurance that Peter was bright and charming was sufficient pedigree. The reality quickly proved how unpredictable was the moon. Peter was a banker, about thirty—harrowing good looks, rigorously sculpted pecs and no love handles, the full body armor. He thought it either cute or original to affect a mock horror at the tack and sleaze of Southern California.

Even in a good year the attitude factor got my blood up: either strain, the northern or the Manhattan, imperious and contemptible but with certain common features—"How do you live without opera?" for instance. It all used to be terrific material for a fight, but now I saw the attitude thing as a form of self-ghettoization, locking us all in plague cells like separate masques of the Red Death. Anyway, this arch kid with the pecs appeared to be unaffected by AIDS. His conversation was full of Wilkes Bashford and radicchio, career and money. I prickled at the sight of him and gave him a wide berth. I realize, of course, that he was trying to be "up" for Craig. He was also diagnosed with ARC within a year and has taken a quiet demotion at the bank, off the fast track. Today I wish him Wilkes shirts in every stripe and color.

Friday night Roger and I went to a Writers Guild screening of *Kiss of the Spider Woman*, where I became so unraveled with grief at the end that I couldn't leave the theater till I'd composed myself. Anything with love and death together was unwatchable. "Too stimulating," as Roger always used to say about the horror movies I dragged him to. The next day I took Craig and Peter to the Getty and almost lost it when I showed them the grave relief of the warrior binding his friend's wound.

Sunday was Rosh Hashanah, and Roger went to the Orthodox home of a fellow lawyer who didn't understand we were lovers. It was one of those once-a-year situations where it didn't seem worth waging the battle, and I wasn't that needful of a new year's dinner. But Bruce had decided to have a Rosh Hashanah dinner, in large part because his roommate, Chana, was willing to pull it together. When Peter left, Craig and I took a walk over to Bruce's and peeked in the window from the bushes. One of the *Manicurist* principals was supposed to be there, and I didn't want to cross paths with him. Fortuitously he had already left, and we headed into the gathering of Bruce's baroque circle of friends, from every angle of show business. We all could have gone in the garage and put on a show at the drop of a hat, like Mickey and Judy.

Bruce was basking in what he did best, bringing people together. I guess by that point everyone there must have known he had AIDS, but I liked being his special pipeline friend, and he was particularly nice to Craig as a fellow warrior. I didn't much think about whether Bruce was looking well or not, though I do remember him holding court after dinner from the sofa. I had so bought into his own enthusiasm about suramin that I didn't think any dark thoughts, or anything final about the holiday. There was a moment later on, however, when Bruce and I were talking, and his friend Jimmy came up and put his arms around Bruce's shoulders. They were only four or five years apart, and suddenly Jimmy looked so much younger, and Bruce so frail.

Before he headed up to bed that night, Craig told me he had to spill a secret that was killing him. For months he'd been referring on the phone to a friend who was diagnosed but didn't want anyone to know, fearful he would lose his job. Now Craig revealed it was Paul Popham, one of the founding members of GMHC, of whom it was said that the death of his lover in 1980 constituted the first recorded AIDS death in New York. Paul was one of the towering AIDS activists; he had a kind of heroic status in New York, with a rectitude and sense of decency that were legion. I'd met him through Craig several times, and now I held Craig as he sobbed with terror and the pain of the secret. He figured he could tell me because I was three thousand miles away.

When Craig left for New York we made a date that he would come back for Christmas, a very remarkable leap into the future, consonant with Cesar's swimming at Thanksgiving. We were learning how to make plans again, and if we neglected to add the zinger—*that is, if we're still alive*—this was because it was understood so well it was time to defy it. Then right after Craig left I was stricken with a bout of diarrhea, so intense that I called the Ferrari doctor. He said that since I was at AIDS risk they had better do tests for cryptosporidiosis and other exotic parasites. Crypto, as we call it, is one of the wasting agents that can halve your body weight in a matter of months. I freaked out as Roger drove me down to the lab to leave stool samples. Afterwards Roger put in a call to Peter Wolfe, who assured me the infection would probably be gone before the tests were back, which proved to be true. But for three days I was terrified that I'd pass it on to Rog. We slept in separate rooms, used separate bathrooms, and I tried not to even breathe near him.

On September 19, when that episode had passed, I had a session with Sam, the central motif of which was: "What if we both get sick?" I brought up the difficult matter of how best to commit suicide when in dire pain. How did one help someone else to die? What if that person got too sick to ask? There were so many stories now of desperately sick men being cared for by lovers who were just a hairsbreadth behind. I had this image that wouldn't go away of Roger and me on the phone, talking between two hospital rooms—me at Cedars Sinai with Ferrari, him at UCLA. Another image had taken root in my mind like a bad whisper years before, something I never wanted to stare full in the face. Roger's mother had had a friend in Chicago who was dying by inches of cancer, in terrible discomfort, everyone just waiting for it to be over. And one day her son went into her room at the hospital, took out a gun and killed her. Where did one summon the wherewithal for that?

If desperate measures had to be taken, said Sam, if I had to find a shot of insulin or a handful of Nembutal to end myself or help Roger to die, I would find the way. As always, I would protect the two of us at all costs and do the right thing; I mustn't worry that I would flinch. The problem was, if I became morose and obsessive about these thoughts I would also destroy the present, where week by week we were finding pleasure in life. In this stark and hyperreal world of the war, I had to focus on our enduring love, for it was every bit as actual as the horror. Meanwhile Sam urged me forward with the conspiracy poems. Writing about AIDS was a small measure of power over the nightmare.

Mr. Appleton returned on the Friday of Dose 18, and we were glad to see him fit enough to go on with the program. At this point I think they were figuring thirty weeks of treatment in the first phase, so Roger was very much in the forefront still. The difficult thing about Appleton was that he went to every seminar and gathering he could find of people with AIDS, and he would go into excruciating detail about everybody's symptoms, how quickly they all careened downhill. Roger and I didn't have it in us to indulge in gallows humor with him about other men's spiraling misfortune. It was easy enough for Appleton, chipper again and still so early. He was the grasshopper, shrugging off doses when he was indisposed, and we were the ants, punctual at nine every Friday, holding the front line. It was as if the Appleton subtext was *Look how well I'm doing—better than so-and-so*. We weren't fooled for a minute. We knew how easy it was to say to the next listener that he was doing better than Roger Horwitz.

That weekend we went to a polo match at the fairgrounds in Griffith Park. The last place anyone would have expected to find us, but yet again the feisty group at Room For Theatre was having a benefit, a picnic at the polo field. As soon as we parked the car and I caught the rank animal smell, the dust swirling with horse shit, my microbe radar went on red alert. I kept imagining the worms and antigens borne aloft on the stable air, kept thinking of Tom Kiwan, now confined to his house because, as Alfred said, "it's hit his brain." When the polo teams came thundering out to limber up, the beasts and the pounding noise frightened me. I couldn't wait to get us out of there.

I think it was all jumbled up with whether or not we should keep the dog. Cats were definitely out, because stories abounded of people with AIDS getting encephalitis from cat feces. You must never empty a litter box without protective gloves. Not that we had a cat, but I'd become quite leery of the neighborhood cats who dozed on our garden fence and switched their tails at Puck. Even the goldfish—Schwartz—was suspect. I told Rog not to change his water anymore, I'd do it. Schwartz in turn was mixed up with a story we'd heard of someone who caught a brain fever from eating too much sushi. There were levels and levels of wrongheaded myths and paranoia, even among us graduate students. Is it any wonder that the ignorant think they can get it from a toilet seat?

Yet all through September I don't recall any worsening of symptoms—no cough, no tenacious fever. Well, there was a slight cough, but nothing more than usual. Besides, we were constantly monitoring and reporting to Cope to be reassured. A cough to clear the throat, we said a hundred times, not deep from the lungs or bringing anything up. But then Roger never coughed at all without my stopping to listen, frozen in midgesture, whatever I was doing. It would be so much less unsettling to say the signs changed over the next two weeks, that the symptoms began to gather toward another bout of pneumocystis. But it wasn't so.

On September 28 we had dinner with the Perloffs and Susan and Robbert in an oddball Polish restaurant, to celebrate Marjorie's birthday, and we were all very merry. I look back on those early-autumn evenings and want to set them down defiantly as evidence of how stable things had grown. Among the shifting veils of magic, this one takes its power from the belief that every lighthearted occasion was proof we had come back to life for good. The full Cinderella version of this illusion was a party being planned by Sheldon for my fortieth birthday. Invitations had gone out to fifty people—Saturday, October 19, black tie, no gifts, to be catered by Trumps. There had been a certain tug-of-war between Sheldon and me about the guest list. He wanted more movie people and power types, while I wanted friends who would find it a hoot to attend a big deal in the ice palace at the top of Bel-Air. Still, I was touched that Sheldon had followed up on his casual offer months before of a party. I even managed a strained laugh at the dark humor of his subsequent remark to Roger.

"How old are *you* going to be this year?" asked Sheldon. "Forty-four? Well, we'll have one for you on your forty-fifth—if you're still here."

October began hot, shimmering with smog. On Wednesday, the

second, Rock Hudson died, about four weeks after his shy and unadorned statement was read at the first AIDS Project Commitment to Life dinner. His death had been imminent all summer, but still it was one of those shocks that said no matter how much money you had, how quality the care, the virus had its own grim timetable. Sheldon called Roger with the Hudson news, and Roger groaned as if a friend had passed away. That same day Bruce phoned to announce a horrific statistic that would soon crop up as gospel in worst-case news accounts. Typically those who'd broken through with PCP lived an average of thirty weeks following diagnosis. Roger had been diagnosed twenty-seven weeks before. We said all the usual things—that the figures had their base in the early years, when so many died at the first onslaught, that IV drug abusers died quicker because they started weaker—but the number thirty burned like sulfur on the white October air.

Then Cesar called from San Francisco to say he was back in the hospital. The cough that had worsened through the summer, the breathlessness as he made his way to outpatient for chemo—he'd finally hit the wall. Yet at first it didn't appear he'd been admitted for PCP. From the sketchy picture he gave me, always trying to minimize, it was his tree stump of a leg that had finally gotten critical. For a week or so it was just a minor hospitalization—for tests, for observation, nothing dire—and Cesar and I talked inanely about what a lovely hospital it was. Nice rooms, nice nurses, all very nice.

On Saturday, October 5, Dose 20, we took it easy and went over to Sheldon's to discuss the birthday menu. Veal chops, we decided. That night we had tickets to an opening at the County Museum for the Cone collection of modern art, and as usual these days I'd crossed it off the calendar because I didn't want us in a crowd. But after supper that evening on the terrace it was wonderfully balmy, with Santa Ana winds, which always either electrify or jangle. On a sudden impulse, we rushed to the museum to see if the crowd had thinned by 9 P.M. When we got there we had the place virtually to ourselves and cavorted among the boiling Matisses, grinning with delight and dragging each other excitedly from canvas to canvas. The attendant documentary material, lush with Left Bank trivia, evoked irresistibly the Paris of the perfect feeling.

The Cone opening is my trump card, my high ace. For Roger was fine that night, completely fine, no illusion. What I hadn't learned yet was the hairline disparity between being fine and being secure. There was the wedge where the nightmare incubated. When we got home from the museum we lay in bed listening to the swirling of the wind in the trees. I called Craig in New York, and he happened to be in a terrific mood himself. He'd met a psychologist during the summer, and things had flowered in the weeks since Craig got back from California. Craig and I laughed carelessly, startled by our own good humor, as if we might have to pinch ourselves before the night was done. Sometimes you manage to bring off a moment so astonishing you can't even say how you did it. You even pretend you can do it again.

Next morning Roger and I went down to Pennyfeathers for a late breakfast of pancakes. We were reading the Sunday paper, Roger leafing through the "Calendar" section, when suddenly his face crumpled. "Oh, no." I looked at him. "John Allison's dead."

There was a picture of John, his smile a Shakespearean imp's, and a moving obit by theater critic Sylvie Drake. She spoke of a call from John during the summer, when he'd said, "I'm in the last stages of AIDS." My emotions were all chaotic—what did he mean by *last*?—but everything fell into place now. That odd talk about giving it all up and going away. What had been his final vague excuse about not going forward with my play? We'll put it off till the fall, he'd said. And Roger and I were so busy with suramin and staying alive that I hadn't ever got back to him. In theater you have to get back to people, keep the energy up. Though we had scarcely known him, we were both blown away by the news. John represented the felicity of life before the moon, as Roger and I recalled the lunch at Trumps, a year ago almost to the day. At the end of my play, when the boy

Tom leaves Julian—Joel is Tom, I am Julian—he asks: "Does it all go too fast?"

"You mean life?" says Julian. "Just the summers."

But where were the symptoms? What was the red flag? All we did was come back for a quiet Sunday, brooding on too many deaths, worrying about Cesar. Monday we went right back to work. I kept an appointment at Paramount, though the only note I have from the meeting is a scribble about John Allison's death. So what was it sent us over to UCLA on Tuesday morning? I can't remember. A fever, I suppose, or the cough in the throat, but nothing out of the ordinary. If the doom was very intense, colliding like ions in the heat-swollen sky, it was only because of all the bad news the previous week. It wasn't us.

Cope must have ordered a blood-gas test, and the oxygen level must have been low, so they decided to admit him for a bronc. Within twelve hours we knew the pneumocystis was back. But all I remember anymore is the bewildering shift of seasons, from laughing among the Matisses Saturday night to the fever three days later and Roger overwhelmed. And they took him off suramin. When we pleaded for them to give it back, they said not while he was on Pentamidine. I remember Gottlieb coming up to me in the fourth-floor corridor. We hadn't bothered to check into the penthouse, thinking we'd be in and out. "I want you to know," said Gottlieb gravely, "we'll do absolutely everything we can." He meant to comfort me, but I just kept beating myself: *How did it get so bad so fast? What did we do wrong*?

Even though I know now that the drug had turned on Roger, I still can't understand how we could have had no warning. Hope had left us so unprepared. We had grown so grateful for little things. Out of nowhere you go from light to dark, from winning to losing, go to sleep murmuring thanks and wake to an endless siren. The honeymoon was over, that much was clear. Now we would learn to borrow time in earnest, day by day, making what brief stays we could against the downward spiral from which all our wasted brothers did not return.

# · VIII ·

Once more Sheldon was there before the night was out, and again he played the single-issue politician: *Don't tell the parents.* I was still trying to find out how the infection could have got past the suramin. I'd made contact with all my antiviral sources, sending an SOS to some, to others a warning. Casualty on the front lines. I was only half there when Sheldon was purring reassurance. No big deal, he said, we already knew the procedure. Get the infection taken care of, and then back to work. The secret was intact; why bother two old people in Chicago when they'd managed to live in ignorance so far?

Roger nodded passively, too sick from all the tests to argue, gearing up for another siege of medication. When I tried to raise the issue that we seemed to have a magic-bullet problem here, and maybe it was time to go after Compound S, Sheldon changed the subject to my birthday dinner, only ten days away. Since the doctors were saying Roger would probably be home by then, there was no reason not to proceed on schedule. It was such a seductive idea, to think we could still breeze in in tuxes and put the calamity on hold. I thought of Bruce at the Oscars in March, nominated for *The Natural*, a moment of tonic gaudiness between the first lesion and the pneumocystis. And here we were, agreeing again to the lie of normalcy and holding out for veal chops.

As to the burden of the secret, it wasn't Roger's parents I was worried about right now. I felt dread enough of hitting my parents with the news, assaulted as they were by the complications of my mother's emphysema. Indeed, we had all we could do, in the wake of the nightmare, to preserve our own dignity about being gay. I don't think we knew what to do yet with our parents' hard-won acceptance, the sense they'd had to overcome that being gay was a kind of doom. So the secret wasn't all Sheldon's idea, even now. We'd protect the parents as long as we could.

But I simply couldn't go on smiling at our friends and coasting along as if nothing were wrong. I couldn't face Alfred in the mornings, or all the calls that were pouring in about the party. I phoned Richard Ide in Washington; he was there for a term's sabbatical and bunking with a mutual friend. I had to banter inanely in order to get to Richard, who in turn was required to speak in coded monosyllables. It just couldn't continue this way or we'd go mad—though now was hardly the time to discuss it. Roger wasn't up to talking to anyone new, and especially kept his distance from the fuss of easy sympathy. I recall how we both looked grimly around at the flowers that had welcomed him home from the *last* hospitalization. "What is this," said Rog with wry dismay, "a funeral?"

But if he didn't need reinforcements, I plainly did. I was berserk, and it was coming out as anger. One afternoon in the underground garage beneath the city of pain, the Jaguar locked in gear again. I came racing up to use the phone in Roger's room, ranting as I dialed and then screaming at the dealer in a sort of free fall of rage. It was a reaction that would soon become a reflex, at every little thing that went wrong in the world of errands and customer service. Pure displacement: I was angry at Roger for being sick.

And it wasn't even being safely funneled off, since Roger had to lie there weak and fevered and listen to the Jaguar rant. "Please, I can't handle all this upheaval," he begged me.

I only wish the yelling had calmed me down, but it didn't. A day or two later he had a call from Tony Smith in Boston, and managed to rise above the fever and nausea to have a quiet talk with his friend. Somehow it made me jealous. *I* couldn't talk to anyone that way now, not in a state of emergency. I don't know what it was I did just then—nagged him to get off the phone, started wailing or getting frantic—but he hung up and turned and shrieked at me. "You can't take it! You just can't take it, can you?"

In eleven years he'd never yelled like that, and this in spite of a lung infection that often left him too exhausted to talk. But he was right; I was going over the edge again. What good was I going to be to either of us if I couldn't take it? And if I couldn't take it now, how would I ever see it through? The savage disdain and loneliness in his cry were as sobering a challenge as either of us ever made. *Don't leave me*, Roger had pleaded with me back in '81, in the aftermath of Joel, when it seemed I didn't know what I wanted anymore. At the time an embrace and a promise were half enough to reassure him; time and a little growing had done the rest. Now I was being asked for much more. Falling apart would just not do.

The first thing I did was tell the truth to Alfred, who sobbed in the Jeep and kept asking what he could do. Nothing, I said, but I knew what I wanted. We'd have to pull back now on work, since I wouldn't be caring about the two deadlines, at CBS and Warner Brothers. There would be no more hungering after career or catering to the hustle of Hollywood. Alfred tried in the weeks ahead to address this issue, suggesting that I had to keep working to keep my sanity. None of it mattered if Roger died, I replied, and when he tried to exhort that I must survive even if Roger *did* die, I distanced myself from him completely. In any case, all I felt like doing, besides keeping watch at Roger's bed and charming a whole new set of nurses, was making my endless phone calls about what had gone wrong with suramin and where the fuck was Compound S. The doctors were being very precise about the current infection, making no connection with the antiviral.

And then, on the third day in the hospital, Peter Wolfe and his colleague David Hardy came whipping into Roger's room. Bruce Weintraub, they said, had just been admitted to a room three doors down the hall. It was extraordinarily sensitive of the two immunologists to care about our secret so. Perhaps because they were near our age, they understood how a young man fights to keep control of the options, for the young still cling to the illusion that their bodies are their own business. Hardy and Wolfe also knew how virulent the gossip could be; both had patients who'd lost their jobs, their friends and their reason. We knew how many familiar faces would be visiting Bruce, back and forth in the corridor. So as sick as Roger was that day, we decided it was time to move to the tenth floor. Since it couldn't be arranged till the next morning, we peeled Roger's name tag off the door, and whenever I left the room I checked to see if the coast was clear.

But though privacy was the immediate issue, part of me was reeling from the coincidence of Bruce and Roger down with infections at the same time. Only in Bruce's case it wasn't PCP, it was something wrong with his blood, a plummeting of his white count, as I recall. This was a whole other territory from the pulmonary department, and my knowledge was sketchy at best. But it struck me then that it wasn't just something hit-or-miss about suramin that had allowed the protozoa to fulminate again in Roger's lungs. There was something more deeply wrong here, something bad that nobody could name yet, against which the elixir was powerless.

I couldn't wait to leave the fourth floor. In just those few days we'd become acutely aware of the man in the next room, obviously gravely ill, his parents in hovering attendance. You pass people in the hall, sometimes walking with their patient while you walk with yours, or in the elevator or the coffee room. Some will spill you the whole story of the loved one's illness, but even the random nods and hellos speak volumes. These two modest parents were clearly bowed down by a very late stage of AIDS. I think there was a lover there as well, and various overdressed friends, but it was the parents you wanted to rock in your arms, they looked so lost. So I was very relieved to flee to the optimistic luxury of the tenth floor.

Once I realized Roger was stable, I forced myself to stay calm. After the shock of the sudden diagnosis wore off, I think neither of us was quite so terrified of PCP as before, knowing he had come through it once. For Rog the harder thing than fear was the disappointment, being thrown down again after climbing a mountain. Familiarity with IV nurses and the protocol of contagion didn't make them easier. But he finally told the truth about what he delicately called "my situation" to one of his law buddies, and thus was able to channel off some of the pressure of work. He was in phone contact with his secretary every day and returned the lion's share of his calls. All of which more or less shamed me into working with Alfred in the afternoons, though in truth I spent most of that time crying and complaining.

I'd stay with Roger in 1016 till eleven or midnight, always enjoying the quiet that descended on the floor late at night. We'd put in long calls to Jaimee and my brother and generally end the day feeling safe. Then on the way down I'd stop by the fourth floor and go into Bruce's room. He was always asleep, but I never had time to pop in during the day. Right now I longed to talk to Bruce about suramin and Compound S. Of all of us, he would have been happiest to sift the evidence for hours and spur me on. But I also didn't mind just sitting there watching him sleep. It grounded me to realize that Roger and I weren't in the fight alone, and Bruce made the hospital seem less overwhelming, more like a satellite station than the moon.

After a while he would stir and wake up and stare at me for a moment in the half-dark. Then he'd flash me a peaked grin: "Hi, Paulie." I don't remember much of what we said: just a few sentences before he drifted back to sleep. He was plainly very ill, and the doctors were stumped. It was sometime during those first days that they decided they'd have to remove his spleen. I remember talking with him about that, very matter-of-fact, toneless in the dark. I also had a very specific memory of the composer in New York who'd had his spleen removed, and all his friends said the operation was "the beginning of the end." And then there was the rumor that any surgery at all seemed to accelerate the AIDS "process," an onrush of final infection.

Yet I'd always hold out for Compound S and swear to Bruce we'd get it somehow. No matter if he was too weak to fight for it now; I would fight for it. Then I would think as I took the elevator down from four that the hospital was going to get fuller and fuller with AIDS cases, till there wouldn't be any beds for anything else. How many AIDS patients would you need before people didn't want to have elective surgery at a hospital, or their babies delivered, or even their blood drawn?

My actual fortieth birthday was Wednesday the sixteenth, three days before the main event at Sheldon's. The doctors were sticking to their promise that Roger would be home in time, but it was equally clear after a week of Pentamidine nausea that he'd never make the party. I don't know why we just didn't cancel it, but Sheldon had a thought. We would simply postpone the party three weeks, till November 10, when my brother and sister-in-law would be visiting from Philadelphia. Bob and Brenda had had to decline the original invitation, and now they would be the excuse to reschedule. Sheldon said he'd take care of all the details, and the fifty guests were called and shifted. Their varying degrees of suspicion about the sudden postponement have filtered back in the two years since, but at the time you grasp at straws. Besides, I really did think we could have Roger on his feet two weeks after the Pentamidine was finished, for that was how we had done it before.

Meanwhile, all day Wednesday, I had to grin and bear phone calls and cards and especially the cheerful wishes of my parents, who had sent forty birthday candles in a soapstone dish and insisted I use them on the fictitious cake I said Roger had brought home. In fact I catalogued for them the whole evening we had planned, dinner and gifts and friends dropping by for cake. Then I went out for a quick supper with Rand Schrader to break the news—one by one, I was convincing Roger to let in our near and dearest. I arrived at 1016 about nine o'clock. We lit one of the two-inch candles and warbled me a happy fortieth. I'd already started a bloody poem called "40," about the final birthday, which took its cue from the losses of World War I. Yet we managed to laugh that night, at the absurd interface of the forty milestone and the fight to the death.

The next night I was sobbing when I came into the main lobby of the hospital. Tears are part of the leeway of the common areas of a hospital, since so many have to do their crying away from the patient's bed. You don't care who sees you cry in the lobby: it was port of entry for all the sorrows, and one gave up all one's previous citizenship at the border. I was tilting across the lobby toward the elevators when I saw dead in front of me a group of five people, friends of Bruce, none of whom I knew well. One, a woman writer, caught my eye and looked shocked. I couldn't pass or duck.

"Paul?" she said, as if she couldn't believe how upset I was. I tried to cover by talking about Bruce, then said I had several friends in the hospital, and after an awkward few seconds they all backed away from my disconnected grief. A few days later it entered the rumor mill that I had AIDS myself. The woman told a friend—another writer—and the story was off, like children playing Telephone.

Somehow we made it home by the end of the week, and by then we thought we knew just what we needed to do. The rebuilding process: the schedule of medications constant as a monastery, work to regain the lost pounds, eat around the nausea, bide the time to go back on suramin. I remember Dennis Cope telling Roger specifically there was no reason he shouldn't regain his strength as before, and he should push himself to get his stamina back. Cope had visited 1016 every evening, not so much to examine as to hear us out and urge us forward. As always, his gentleness of manner and quiet optimism occasioned the deepest healing in us. He was encouraged about what we were hearing of Compound S, but noncommittal about whether it was an option for Rog. Perhaps he was having doubts himself about how much time suramin had bought us.

So we tried to do just as we were told. Susan and Robbert brought over a cake on Saturday, since they were party to the secret now. Though the stated point was not to let the black-tie night pass uncelebrated, in fact it was Roger's homecoming we were cheering. He was still dozing in the bedroom and Susan and I were setting the table, when the doorbell rang. I opened the door without even thinking, and there was Bill Ingoldsby, all spiffed up in a tux and carrying a bouquet of flowers about four feet across. Sheldon's secretary had skipped his name on the recall list, and he and his friend Dan had been driving around the darkness of Sheldon's house, wondering what had gone wrong.

I stepped all over myself apologizing, but not wanting to invite him in either. I hastily explained about the change of plans, then steered him down to his car, where Dan was waiting in a new tux he'd bought for the party. I stood holding the great floral tribute, spinning off lies—Roger was out for the evening, I had a business meeting going on, everything was fine. Regrettably they were going to be away on November 10, but they couldn't have been more gracious, relieved to hear nothing was wrong. As they drove away I was racked with guilt and inadequacy, longing now for everyone to know, loathing what was left of the charade.

I have virtually no record of the next three months. Except for a few doctors' appointments, Roger's calendar is completely blank for the rest of the year, and he wouldn't even bother with a calendar for '86. Between then and the end of January there is a single five-line entry in my journal, and my daily calendar is as empty as Roger's, because I ceased to write my appointments down. I kept the ones I could remember. Indeed, we both went on working as long as we could, struggling into November, but it was as if the whole idea of calendars had become a horrible mockery.

I wish I had an account of just the meals we ate, or a log of the calls that came in, for there was where we lived. From now on we wouldn't be spending much time in the abstract, not at least as it related to future or careers. Besides, when you live so utterly in the present, the yearning to record it goes away. To write in a diary you have to hope to read it later—or to last long enough to make the appointment two weeks down the road. Right now you are trying not to vomit dinner.

I remember the whole of that autumn as ominous and desperate, but that had as much to do with the hurricane of other cases as our own. I was talking to Cesar now every third or fourth day, but except for my telling him I loved him I don't remember what we said. He did finally get diagnosed with PCP, and was on Pentamidine at the same time Roger was. He would tell me how one old friend whom we'd always found maddening would come in the afternoons and read to him, thus wiping out all his black marks.

Most precious of all was Dennis, who'd arrive at the hospital every evening directly from work and was, in Cesar's oft-repeated phrase, "an angel." If I remember nothing else about Cesar's last weeks, I recall the opulent tenderness of his feelings about Dennis. So intensely had he lived the Platonic intimacy of the last few months that it flooded his mind with light. His voice grew fainter and fainter, more and more tired, yet still he could laugh coquettishly or brag that all the nurses and the other patients on the floor knew Dennis was his special friend. Sometimes I would talk to Dennis himself in passing, just for a few moments, and he would make oblique remarks about how bad it was getting. Yet with all that, I knew Cesar's friends were trying to arrange for attendant care at home so he could leave the hospital. I didn't think of it as a hospice situation. I couldn't really think that far ahead, or perhaps I couldn't bear to.

And for some reason nothing was ever said about me visiting. I guess it just wasn't an option because I was so busy taking care of Rog, and I don't doubt Cesar leaped to defend me to anyone who wondered where I was, his best friend for seventeen years. Or perhaps he simply told them all the truth about Roger's situation—he knew the secret was over now. But though I've made my peace with not going up to say good-bye, I wish I'd been able to talk to him to better effect than I did. For it seems to me I kept promising him drugs he was clearly beyond the reach of, and silences would develop on the phone because I couldn't laugh or think of a witty retort. Maybe it was enough that we kept on saying we loved each other. That is all you are sure of afterwards.

"Hello, darling," he'd say when he heard my voice, his own voice sweet and grinning as ever, no matter how faint. And then before we hung up: "You keep the pool open. I'm coming down for that swim."

Roger's recovery proved to be discouragingly slow, and the nausea that went with the Pentamidine seemed worse this time, so he wasn't eating well at all. I'd had him on a regimen of vitamins during the so-called honeymoon, but now his stomach was too queasy for him to take the pills. Once when I insisted, he choked on a mineral capsule and heaved up half a day's food, sending me into a wave of hysteria. One likes to think one will be endlessly gentle, easing the difficult symptoms, always comforting, making light of every indignity. But the fear and the heartbreak twist you up, and your own helplessness blinds you till you don't even take the modest steps you can. "Hysteria is not sexy," as Cesar once said in another context. Soon you are absolutely fixated on every meal, for that is still the best you can do, and when it's not good enough you start banging pots in the kitchen and stuffing whole meals down the disposal.

The one person who could calm me down and make me see the minor crises in perspective was Roger—the only one ever in my life. Over the years, relations between us had evolved to a place where he was the grown-up and I the child, at least in matters that required the filling out of forms, lines to stand in, the engine of running a house. Roger always seemed to take care of everything, and now that state of affairs was in flux, because he simply didn't have the energy anymore. He who had always been so independent, who'd lived on thirty dollars a week in Paris, now had to sit and be waited on while he recovered, with all the attendant hovering. He'd hand me bills to pay, and I'd go bananas trying to keep the seven accounts straight. Not the least bit sexy.

Yet we would take our stamina walk up Harold Way in the late afternoon, and Roger would say in anguish, "I don't want to be an invalid, I don't want you to have to take care of me." And I would fire him with a speech about our interdependence, gripping his shoulders as if I would fuse us into a unit. The minefield of lunch would be forgotten, the byzantine mess of bills. We still had a feel for loftiness, and there was only one way to go: onward.

The one perspective I did seem to have was that Roger was doing better than Cesar or Bruce, and I told myself over and over to worry about them instead and fight to keep them in the arena. One afternoon I went over to UCLA to pick up a load of medication at the pharmacy—something for the nausea, plus an oral dose of Pentamidine, which they thought might be prophylactic against the PCP. Bruce had had his splenectomy several days before and had come through it fine, so I stopped up to visit him. His parents were there, his mother bewildered and knitting. Bruce was in antic good spirits, with two or three friends around him and his sister Carol calling in from the East. He made me laugh and treated me like some kind of special envoy. Then his lawyer came in with his will to sign, and the rest of us repaired to the waiting room, where Chana, his roommate, held my hand—for my sake, for hers, who knew anymore? I went into an automatic lecture about Compound S, and I remember Bruce's father listening hungrily.

Checking in with Bruce and Cesar was my way of assuring myself there would be no break in the line, for they were my platoon. Our deep-pocket source in the UC system was already beginning to bend elbows about Compound S, so we all just had to hold on. If Roger could be home and fighting after two bouts of PCP, surely Cesar would be fine after one. With Bruce so irrepressible on the day of the will signing, it was easy enough to believe the short-term notion of the doctors that a person could live as normal as you please without a spleen. A person with normal immunity, they might have said. I'd bring all the reports home to Roger, and we'd hold tight to our little population sample, relishing the safety in numbers.

By the next week, on sheer willpower, Roger started going into the office. He was still feeling dreadful and queasy, and the doctors couldn't quite understand why he wasn't bouncing back. But weak as he was, the very mobility seemed to prove we were on our way again. I recall an afternoon when I took over a poetry class for Carol Muske at USC: I had to teach William Carlos Williams's "Queen Anne's Lace," which put me in a swooning mood. After the class I called Roger at home. He was all excited because he'd just had a new bulletin from a friend about Compound S and the extraordinary results they seemed to be getting at Duke, especially from those who went on the drug early. "This one might really work," said Rog in a kind of stunned delight. I started to cry with relief. I was so giddy with hope I could barely drive home.

We heard we might have to go to North Carolina ourselves to get the drug, which now we knew by its clinical name—azidothymidine, or AZT. Months before, I'd had the vision of Paris and the barracks of HPA-23, everything changed by the war we bore within us. Now

I imagined a mild autumn in the Great Smokies, a hospital terrace looking out on a view that would taunt us with loveliness, like *The Magic Mountain*. Then a few days later our source told us we might be able to get the drug in California after all, perhaps within the month, but it would all have to be top secret. The thrill of the undercover operation kept us going, and this at a time when AZT had the status of a Holy Grail in the AIDS underground. I'd immediately pass on all the details to Bruce and Cesar to keep *them* going, then call Craig in New York. Don't tell anyone, I'd tell everyone, meaning shout it from the rooftops but you don't know where you heard it. There was a time when I must have been the major leak on AZT.

No doubt we were letting ourselves get lost in magic again, just as we had with suramin, but with one important qualifier: this time the silver bullet was for real. Yet we almost lost the war waiting for the bullet. Maybe we should have been demanding something more immediate from the doctors, to get to the bottom of Roger's wearying symptoms. But I'm not sure they would have caught the problem, even with a lot of probing. The general AIDS symptoms are so diffuse and fluish that they could be evidence of anything or nothing. More often than not, you just have to wait and see. And it's all so changeable: already we were experts at the triage of good days and bad days, where you develop a hundred coping mechanisms to get you through. Late at night we'd be lying quietly in bed together, reading and watching the news. Always the hope that the next day might be better, and only a few more weeks to the next elixir.

Meanwhile our short-term goal was the party on November 10, and the visit from my brother and sister-in-law. Somehow these two bustling events mitigated or even masked the bad days, the fevers and fatigues, though finally Roger began to reconsider about his parents. Sheldon and Jaimee were both still arguing that it wasn't necessary to bring the parents in, but Sheldon didn't even know Jaimee knew, and Jaimee in Chicago couldn't see how overwhelmed we were. She was already possessed by the hope of AZT, as fervent as we were.

As November began, the only concrete detail I can recall—beyond the dinner battles and the assaults of malaise—was the matter of the Ganges. The house just above us on Kings Road had a problematic septic tank, which would now and then kick up and overflow. Then a rank-smelling stream would run down the street along the curb by our house. It stunk like raw sewage, but the problem was intermittent enough that we'd ignored it for years. Cesar had caught a whiff of it long ago and dubbed it the Ganges. Now for some reason the septic tank broke down in earnest, and the Ganges was flowing every day, sending up a stench we could smell on the front terrace.

I wrote a polite note to the neighbors, but nothing happened. I lapsed into an exponential terror about the infections carried by raw sewage: I realized the dog walked through it every day, then sometimes jumped up to curl at the foot of the bed when Roger was resting. I had already reached new heights of cleaning, my rag streaming with ammonia nightly as I wiped every surface. I'd throw away half of every lettuce and wash fruits till they whimpered. When the Ganges was flowing I wouldn't even let Roger have lunch on the front terrace, for fear of airborne horrors. I finally confronted the septic neighbors and said Roger was sick—sick, period, not the "A" word—and they simply had to do something. They were two gay men, both psychologists, both working heavily with AIDS patients, and at last they hustled to get the job done, though it took till after Christmas.

That first week of November, Alfred came over with the news that one of the cases we'd been following had shot himself dead the night before. This was a man of some wealth and prestige, very big in the art world, and he'd walled himself in with his secret. Alfred and I talked with the strange dispassion we have these days about whether the guy had been too sick to get pills, or had his friends all refused to make it easy? Did somebody bring him a gun, or did he have it already? It was then I made Alfred promise that if I got to that place and no one else would get me what I needed, he would provide. He said yes, but there's no way of knowing if he meant it. You never know till you're at the wire. A friend of mine in New York promised a friend of his to help him die, but when the time came the dying man was two thousand miles away, with a family who loathed his gayness and the sin of his illness. Luckily, he was lucid enough to pull his own plug.

Three days before Bob and Brenda were to arrive, my brother called to say he had a kidney infection. This necessitated a worried call to Cope, who reassured us Roger wouldn't be in any danger, but I was tense and frightened when they first arrived. My brother remembers how manic I was, scarcely able to sit for five minutes without going in to check on Rog. He says Roger was very thin—not Auschwitz-thin; that is a different stage entirely—and he recalls the difficulty Roger had eating and keeping down food. On Saturday we all went out for lunch, but Rog scarcely touched his meal, and had to excuse himself to lie down in the car.

Saturday was the tenth, the night of the black-tie dinner. In the afternoon Bob and Brenda and I stopped by Sheldon's house in Bel-Air to check on some last-minute seating arrangements, and Sheldon and I had a brief talk upstairs. I think until that day he kept hoping Roger would pull it out of the hat and appear for dinner. Sheldon seemed to want to say something to me, but he kept veering off into tangents of humor. When I talked about what a hard time his brother was having, I remember him saying his grandmother used to tell him, "Anything's better than being dead. It's better to be a whipped dog than dead."

It was pouring that night, and Roger took pictures of the three of us in our evening gear and sent us off with the admonition to be good children. All the way over in the car, we strategized about how we'd handle the schizoid nature of the evening. Perhaps fifteen of the fifty knew the truth, and another dozen suspected as much and were waiting for us to make a move. But Sheldon had won his point on the guest list weeks before, so there were twenty more who were black-tie regulars, perfectly nice acquaintances from the business who would presumably buy the ridiculous excuse that Roger was down with the flu.

Sheldon's house is designed for parties, the main room forty by forty by twenty, all in sand tones, with ceiling beams that are bleached telephone poles—no art, no color, the whole design vaguely Santa Fe but wildly outscale and utterly un-Indian. I went around with a frozen smile, introducing my brother and sister-in-law. My brother, because he was in a wheelchair, was perhaps sufficient decoy for people to cover their clumsy feelings about where Roger was. I spotted a couple of people and grimaced at the gauntlet of chitchat I would have to run. As soon as I escaped into the dining room to see about the seating, I placed a quick call to Roger from the kitchen to get some moral support. I'd be fine, he said, dozing comfortably. I should just try to have a good time.

Sheldon ran a dinner as smooth and punctual as a board meeting. The flow from the bar to the dining room was terrific, the catering top-drawer. Bob and Brenda and I sat at the center table with Sheldon, his half-brothers and their wives. Roger's cousin Merle was down from San Francisco, and I put her at a head table with friends—including Rand, whom she'd dated at Berkeley decades ago—because she was still reeling from the news, broken to her only tonight. All during dinner I wrestled with whether to read my "40" poem. I'd mentioned it to Sheldon that afternoon, and he'd nixed the idea. He didn't have a clue it was all about AIDS and the end of us all, but I think he had an instinct that it wouldn't be about the moon in June.

Between the veal and the salad, I slipped into the breakfast room and called home again. "Should I read my poem, Rog?" I'd finished it only a few days before, and no one had heard it but Roger and my brother.

"You do what you want, darling," he said mildly. "It's your birthday."

Then Sheldon popped his head in. "What are you doing? You've got fifty guests out here."

Sheldon gave the first toast, which was warm and oddly intimate. He said he'd tried to tell me that afternoon that he loved me, but couldn't find the words. It is the truest thing about him that he could only say it now, in front of fifty people, but it was no less moving for that. He had earlier gone around and fingered various friends to follow him, choreographing a seemingly spontaneous outpouring. They all rose in turn, in the proper order, but there Sheldon's control of the flow of events ended.

The toasts were uniformly sober, not very successfully couched and extremely painful. Susan toasted Roger and me, saying what great friends we were. Charlie's voice cracked as he toasted "my most original friend, who's always there for all of us." Carol spoke about our poetry conspiracy and ended with: "You've taught us what it means to love."

It's not that I'm so wonderful. These very friends have seen the wallpaper curl from my overwrought opinions. I am nice to old ladies and dogs, but otherwise I might say anything at all, often about as subtle as a pipe bomb. But they were all having the same problem I was, staggering under the subtext of this party, a tragedy all around it like the Alaska storm that roared across the high Bel-Air hills. So they spoke their valedictories to a life that ought to have been an Astaire and Rogers movie by the time it played Bel-Air. And everyone wanted to spill a drop for the absent friend who anchored us.

I don't know what the twenty still in the dark were thinking by that point, but when I stood up, all I wanted to do was say it out loud at last. I knew then I had called home to get Roger's permission. First I volleyed the compliment back to Sheldon and said I loved him too. Then I said I'd always thought turning forty would matter, and here it was and it didn't. I said I wished Roger were here, and then

I began the poetry lesson, giving the whole background for the conspiracy to a group half of whom probably hadn't cracked a poem since high school.

The poem is all about dying at forty, and its main figure is Robert Louis Stevenson, dead in Samoa at forty-four. The poem is forty lines, and after the bit about the wasted generation of World War I, it goes on with the nerve-racking business of waiting to die:

> *ask any phobic it's not the heights*
> *it's the edges that get you that weird thing*
> *of being drawn to the precipice do it*
> *the time has come to take the plunge and none*
> *of your youthly coy and basket shots will*
> *save you time doesn't give a fuck oh but*
> *we planned such plans if the war hadn't come*
> *and the weather had held and life had cleared*
> *like a late Manet . . .*

Stevenson died half a world away from his native Scotland. Now here we were, a world away from being young, "bone-thin and sunburned, blown like a sailor." The poem ends with a wish windy enough to blow out forty candles, a wish that in the end was granted:

> *. . . if I*
> *must go early give me please one friend one*
> *year but nothing's enough and the cliffs at Thera*
> *where the old world ended tomorrow my love*
> *is a stolen kiss but we sail together*
> *if we sail at all hey 40's kid stuff.*

Sheldon gave the waiters a little nod, and they came around with little icy wedges of chocolate cake drenched in raspberry sauce. I sat down again with the family, and one of Sheldon's sisters-in-law observed that she hadn't understood a word of the poem, and one of the half-brothers grinned at me and said, "Not a lot of money in poetry, I'll bet."

We were all out of there punctually by midnight, the catered syncopation never wavering. Bob and Brenda and I swept home through the rain, talking nonstop, and woke Roger up with the details. My brother reminds me the next day was a good day. Roger was up and around, animated with Bob and Brenda and loving the closeness of family. Good days are such a mysterious gift that you dare not question them much, and the only problem is they give you a false sense of security. That night we had dinner at Cock 'n' Bull, and Roger put away a plate of prime rib, leaving us all daft with merriment. When we got home there was a message from Ted Hayward in San Francisco, a friend of Cesar's and mine who'd been very close to the case. I thought Ted must be calling to give me birthday wishes from Cesar. When I phoned the hospital and was told Cesar wasn't there, I assumed they'd negotiated the move home. Denial doesn't get much deeper than this, but please, it was a good day.

Monday wasn't. Roger was feeling awful and could barely get out of bed to come into the living room when Merle dropped over with Sheldon to say hello. There was a long discussion, I remember, about telling Roger's parents. I knew Roger wanted to now, and when Sheldon made a last attempt to preserve the secret, I shifted the conversation to "what Roger wants to do." Eventually we got a consensus that it would have to be Roger's decision. Then I took Bob and Brenda down to The Source for lunch, feeling a wave of irrational guilt that the clouds and rain were so persistent. As we waited for the food, I went over to the pay phone and tried Cesar's number again. Ted answered.

"Between a flash of lightning and a clap of thunder," he said when he heard my voice.

"What? Is he home?"

"Saturday night, just after midnight."

I went totally blank. Then I groaned with frustration—no, annoyance—and said: "Oh Christ, how am I going to tell Roger?"

Cesar died on my black-tie birthday, within an hour of the poem and in the thick of the same thunderstorm. I don't recall any of the details after that, except that Dennis had had to go out of town, and by the time he landed in the Midwest there was a message summoning him back to San Francisco. I got off the phone as quickly as I could. As I walked across the restaurant, the seventeen years of friendship ended. I told Bob and Brenda, who were starting to look numb from all the ravages of war. We went home, I told Rog, he cried, I didn't. I never cried for Cesar.

But I tried to be there for Roger about it, because he took it hard. I remember him saying in the car one night, "Is he going back home to Uruguay?" It was the first time I ever thought of the separateness of the remains and how they could get lost again in Uruguay, the place from which Cesar had finally escaped. Bob and Brenda left on Tuesday, and I began a round of condolence phone-calls—to Dennis, to Jerry, to Diana Cobbold in Massachusetts. Dennis said he felt as if he'd lost a lover, then added that he'd have liked to pour the final urinal over the doctor's head. I reminded Jerry of a dinner party at his house five years earlier; he and I were the only two of the seven in attendance who didn't have AIDS yet. Diana, who'd introduced me and Cesar and was writing a novel about him, was as stupefied as I was. "He never had his great love," she said, and I thought: At least he had the beginning of it.

How sick was Roger that week? I don't know that I noticed anything very different. We were still struggling to hold his weight—he'd lost six or seven pounds now—and I kept taking comfort in the thought that there were people who'd lost sixty and eighty pounds. He managed to work that week too, but fewer and fewer hours. The symptoms—nausea, lassitude—remained stubbornly nonspecific. The AZT was on the way; it would be available within two weeks. If all you have to do is hold on, you let the details go.

Midweek I had a call from Chana. She was on her way out to a meeting, and there was no one to stay with Bruce. Could I please come over? She was frantic. It was only a block away and only for an hour, so I said yes, even though I hated giving up my late afternoon with Rog. As I trotted over to Hedges Place I didn't quite understand why someone had to be with Bruce all the time. Didn't he just sleep a lot, like Roger?

The air was eucalyptus sharp after the storm, the view from Bruce's terrace clear to Catalina. Alpha Betty Olson, a writer friend of his, had dropped by unexpectedly and said Bruce was in his bedroom and would be right out. She and I laughed to think how eagerly Bruce had always wanted the two of us to meet. "He gets his way eventually," she said. Then I raided the kitchen for nosh, because Bruce had a lot of friends who brought up very high-toned takeout.

Suddenly he appeared from the bedroom end of the house in a long robe, stamping in in a fury. He seemed to be mad at everything, but for a moment I couldn't take it in because he looked so awful, drained and thin and frail, much worse than when I'd seen him at the hospital two weeks before. He was angry at Chana for leaving for her meeting. Then he segued into a great rage against doctors, till he had to lie down on the sofa, exhausted by his own upheaval. Alpha Betty and I tried to engage him about one thing and another, but it was the sole occasion when he didn't want to hear about AZT or anything else positive, so I just shut up. And thought: Why am I here and taking all this abuse, when Roger's waiting at home and wants me there?

But something else was going on—something was slightly off center about this fit of anger, as if Bruce himself had gotten lost in the fire. I wondered if they were worried that he'd kill himself, and was that why he shouldn't be alone? Or was there some kind of viral static in the brain? Nothing scared me more than the brain. When I got up to leave, Bruce had calmed down, and he said, wearily but himself again, "I'm glad you came, Paulie. I'll be better next time." I never saw him after that.

Did things get worse and worse that week? I suppose they must have, but Cesar was dead and Bruce was in terrible shape, so worse compared to what? Roger's parents were the next hurdle. The night he called them in Chicago, I didn't want to be in the bedroom with him. Did I think he wanted the privacy, or was I afraid we'd be punished at last? I hovered in the hallway, dreading to hear the tears, always thrown when the stoic lost it. "I'm not getting better," he said, and then he started to cry.

"Do you have it, son?" his father asked gently.

*It* had always been on their minds, though they'd wished it away when they saw us in August. They told him how much they loved him and said they'd be out the next week. Once Roger had told Al and Bernice, it was my parents' turn. My brother and I agreed it would be better for him to break the news in person, so he and Brenda drove up to Boston three days after they got back from California. My mother said she'd suspected something was wrong ever since we left in August—not that Roger had looked sick, but when I'd call to check in on Saturday nights she'd ask where we were going, and I'd say, "Oh, we're just staying in tonight." Curious, the inadvertent clues you leave. My parents had lived our Saturday nights vicariously for years. Now they were generous and supportive, telling Roger he was like a son to them. By the end of the week we had both shaken families on our side.

Not seemingly such a big deal, unless you have heard all the stories from the other side. Craig's mother cut him off one night as he complained about the blood tests and the circular doctors' appointments: "Listen, this whole thing is your own fault. I don't really want to hear about it." That turns out to be rather mild, and at least it's honest. The real hell is the family sitting in green suburbia while the wasting son shuttles from friend to friend in a distant place, unembraced and disowned until the will is ready to be contested. And even that is to be preferred to the worst of all, being deported back to the flat earth of a rural fundamentalist family, who spit their hate with folded hands, transfigured by the justice of their bumper-sticker God.

Either the symptoms didn't seem to be getting worse, or they took second place to the drama of our parents' arrival on the moon. Saturday, November 17, was the Gay Community Center dinner at the Beverly Hilton, and I decided to go and host the table Roger had put together from his hospital bed the previous month. In the Hilton I ran into Rick Honeycutt, no longer impish and surferlike, looking tired and old as he told me he was off suramin. Eagerly I gushed about AZT, but he didn't want to hear about it. Several people came by the table and asked where Roger was, none of them having any idea that he had AIDS, and I said defiantly that he was doing fine and waiting for AZT. Nobody knew what the acronym meant, but they got the picture.

Charlie Milhaupt drove me home, and we went in to see Rog, who woke up in a smiling mood and said, "I just had a dream. There was this green liquid. And all I had to do was drink sixteen cups of it and I'd be fine."

The elixir dream. We laughed at the lovely fantasy of it and went to bed that night with nothing more on our minds for the week ahead than awaiting his parents' arrival and then the drug. We'd given both families a full measure of hope when we broke the news, for we had our pharmaceutical ace in the hole. But Sunday it was glaringly clear that we couldn't just sit and wait. The nausea was intensifying, the fevers were back and the fatigue had reached a stage where Roger could hardly get out of bed. We went over to UCLA for tests, and they admitted Roger, again just overnight, but by now we knew what a euphemism "overnight" could be.

And suddenly my memory is as blank as my calendar for almost the whole of the next two weeks. I remember only the bitter disappointment, to think that Al and Bernice would have to find us in the hospital. I know they arrived on Monday night, four days before Roger's forty-fourth birthday, by which point we knew he had hepatitis, of the type called NON-A/NON-B, noninfectious and probably drug-related. That would explain the nausea and lethargy, as well as the sunburned cast of Roger's face, which began to take on a dull gold flatness.

But the real point is that he nearly died that week—closer than he ever came in the whole nineteen months—and I don't even know when. You'd think the shadow of death would have your nerves screaming to imprint it. Richard Ide says he talked to Roger from Washington on Sunday night, soon after we checked in, and Roger was terrified and started to cry. "I love you, Richard," he said. And Richard knew in that instant that Roger was dying, that this was a call to say good-bye. Scrambling, Richard said he'd be on a plane and be in L.A. by Friday, so Roger had better hang on till then. Yet the fever point of the crisis had apparently passed by the time Richard landed on Friday night, so it must have been Tuesday or Wednesday Roger almost died.

But what exactly does "almost" mean? It wasn't until the next summer that I could even admit how bad it had been during the days of late November. At the time, anything anyone said about dying, however veiled, I simply didn't hear. Because he *couldn't* die, not with the drug just a week away. For this was precisely what was so tantalizing in the rumors of AZT, that it was turning people around even from the verge of nothingness. I recall wanting everyone to let us alone with the hepatitis—no treatment for that but time, the doctors said. We would take care of the time. Just get us the fucking elixir.

Not that our deep-throat sources weren't moving heaven and earth to acquire it. Word was that thirteen or fourteen patients were on it now, but every single one was back east and close to the National Institutes of Health. Superpower threats had to be made to coax the drug to California, and even so the manufacturer had all the time in the world. Meanwhile I had to deal with a pugnacious, cocky little intern called Runyon, barely five foot seven, who wouldn't stop running tests because he wasn't satisfied with hepatitis.

I grew maddened with all of Runyon's probing, but he managed to convince Dennis Cope that they ought to go one step further. Now came the first spinal tap, the first bone marrow biopsy, both tests as awful to contemplate—much less undergo—as medieval tortures. I remember Roger curling up in fear of the marrow test, holding my hand as Runyon, utterly lacking in bedside manner, explained the procedure in ghoulish detail. Yet where would we be without Runyon, bless him, who wouldn't stop brooding over certain ambiguous numbers, and who finally figured it out: Roger's adrenal glands were failing. I don't even remember why that is fatal; I only know it's treatable. When you live on the moon, *treatable* gets to be the holiest word in the language.

How Runyon crowed with triumph! He was easily five foot nine by the end of the week. And within an hour of his diagnosis Roger was on medication—Florinef, a terrific little lilac pill, one a day for the rest of his life—that would do all the adrenal functioning that needed doing. It wasn't until a week or two later that reports began to filter in through the AIDS underground that four suramin patients out of a hundred had lost adrenal function. So it was just an unlucky side effect, that grim companion of healing.

As for the suramin—water under the bridge, which seemed more lethal with every report that came in—of course I anguished to think how much we had wasted on snake oil. There's a moment in *Sunrise at Campobello* when the Roosevelts have been tirelessly giving some vigorous treatment to Franklin—rubbing his legs for hours for the circulation—and the doctor tells them they've been doing precisely the wrong thing. The sinking feeling is indescribable as you reach the dead end and realize you can't even go back to the fork in the road where you took the wrong turn. I felt ridiculous and ashamed, I who had pushed suramin all summer as practically a miracle cure.

But if I was gullible, there were others who knew exactly what they were doing. Though UCLA quickly moved to dismantle its suramin study as soon as it became clear the drug was too toxic, several other suramin programs were still going full force, with hundreds of patients clamoring to get in. Within a few weeks this moral blurring to protect the experimental data began to seem criminal to me. There was one doctor who kept his patients on suramin through the winter, even when we knew how lethal the side effects were, and even as the patients died off one by one.

It is mostly myself that I can't remember that week, perhaps because my panic wasn't manifesting as anger or depression. It was all fear, pure as oxygen off a line. Al and Bernice were there, of course, and they tried very hard to defer to me. Roger and I would talk through the day's numbers every night with Cope, who would calm us down merely by talking about it all as an ongoing process. Now and then Al and Bernice would join us for that session, and once Bernice went into a state of suppressed rage—exploded later in the corridor—when I talked with Cope in front of Roger about the chances that some horror or another might develop. Bernice came from the school that didn't talk about the dire stuff in front of a sick person. In retrospect I agree—Roger didn't need more gloom and doom—but at the time I felt we had to go through the fire together, that all we had to squeak us through was the fact that we were one. I remember Cope taking me aside one night and saying carefully, "This is the most unstable I've ever seen Roger." Then he asked if Roger and I had ever talked about life support systems and a living will. Still I would not hear the knell of death in all this. I think I thought I could disbelieve it away.

In any case, I was a better combatant that week than I was an observer. Fifteen hours a day I'd either be on those interns and nurses like a rash or be plugged into my sources all over the country, wired for sound. Perhaps Roger is the better witness here. Most of what I know of the blackness came out of a long talk we had, late one night the following summer, when things were quiet. Those were the nights when I used to read Plato aloud, and Roger could barely see.

"Oh, Paul," he said, "I had to fight so hard to keep from going under."

He remembers his father joking and his mother giving him foot massages. They'd stay with him all day long, from breakfast till dinnertime, and I would come over in the afternoon and begin my dogged work as an intern without portfolio. Then at seven his parents would leave, and I'd stay till after midnight. Roger remembers us all trying to talk normally—talking mostly over his head, but including him too, as if we were four around a table. How he would cling to that ordinariness, he said, as he held on minute by minute.

It is such a curious business, how not to be alone. In one way Roger was very far from us for days, tucked half in a ball as he dealt with the waves of nausea, the difficult breathing, the general air of being under water. He couldn't even separate the hepatitis symptoms from the adrenal failure, let alone the underlying viral symptoms: they hit him like an earthquake, a typhoon, an eclipse, all at the same time. He counted how long he could hold out by how clearly he was following the conversation from the real world—Al and Bernice and I gabbing inanely about everyone we knew—versus the hospital world of trays and vital signs.

The countdown to AZT was five days, four days, and I would tell him what new shred of evidence I'd heard about the drug. We kept talking to Rog and getting little answers, a weary yes or no, or just a nod on the pillow. We were all so close and so alert, like a troop of sentries. I've never before experienced the feeling of having to physically keep Death away, as if he would actually come in the door if I let down my guard for an instant.

And sometimes you win. Jaimee remembers Roger's birthday as a happy occasion, friends dropping by with presents and Richard Ide in from Washington. Jaimee herself had sent a huge box of food to fatten Roger up, and I unpacked it all over the bed as Roger laughed with her on the phone to Chicago. Of course she couldn't see how bad he looked from his ordeal. Richard recalls being shocked at how thin and battered he was. But mostly I remember—so did Roger—his father telling people for days after, chuckling with a kind of delirious relief, "I really thought we lost him there for a while."

Everything didn't get better right away. He wouldn't be home till mid-December, and he had another infection still to battle. Arsenals of medication had to be consumed. But by then we were all strong, and we weren't going to lose. For on Tuesday the twenty-sixth, two days before Thanksgiving, the elixir arrived. Roger was put on an IV dose of AZT every four hours, the first person west of the Mississippi. We were very grand from that day on, I dare say, or I was anyway. I felt as if we had won the Nobel Prize for our work in immunology. The whole AIDS underground cocked its ear to the tenth floor at UCLA, as a hundred skeptical doctors wandered in and out to get the scoop on the latest magic bullet. Just call us Command Central.

The one note in my journal is for Thanksgiving night, three-thirty in the morning, a lush Somerset Maugham rain beating against the windows: "R Day 3 on Compound S. I massage him and he says wonderful things about me." This was blackmail, actually. I was not half so patient as Roger's mother about giving massages—she'd knead his feet for a half hour. I'd get tired after five minutes, bored more than anything, since the massage would always put him to sleep. But then he'd stir and complain, "Don't stop." That night I said, "I'll keep doing it if you tell me how much you love me." So I got him to purr endearments at me—"I love you so much, you're my best friend"—while I worked his muscles. If he was quiet too long I'd tap him and say, "More." Notice that no one looks over his shoulder to see who might sneak in. Everyone's getting exactly what he wants. And I marked the holiday thus, a pilgrim's prayer in a new world, repeated over and over: *Thank-you for Compound S.*

O bountiful land.

# ·IX·

By December 1, Roger's third week in the hospital, his weight had dropped to 130. It was the most emaciated he ever got: only ten or fifteen pounds below his normal weight, yet his face was so thin you could see the skull beneath. He was exhausted and shell-shocked from the punishment of the previous month, but on the other hand, the hepatitis symptoms were subsiding. He could eat again, and the tenth floor is terrific if you want a lot of food. You can *à la carte* the high-caloric stuff till it fills the tray, and they'll whip up a milk shake on five minutes' notice. Roger's father chimed in with me as we coaxed Roger to snack between meals. On top of which we had him on three cans of nutritional supplement daily. Every night I'd wheel in the upright scale for the weigh-in, and steadily the pounds crept back.

Gradually we all released ourselves from the white-knuckle grip of the previous month. Bernice sat in the chair by the window day after day, knitting a pearl-white sweater for Rog, a sweater she later admitted to me she didn't care if he ever wore or liked; she was knitting to keep him alive. Meanwhile Al joked and flirted with the nurses, as easygoing as he must have been with a hundred different regulars at the H & H. He possessed a wonderful breeziness of spirit that was infectious, plus a gut respect for how hard people worked. And he didn't miss anybody. After a couple of weeks I realized he and Bernice had a chatting acquaintance with Lily, the black woman who swabbed the bathrooms, and Clarence, who did the heavy maintenance work in the corridors.

I was on nodding acquaintance with both, but Al would ask Lily every day about her grandchildren. Always a cheery hello and a bit of sly banter, no matter who came in to take Roger's temperature or deliver his tray. Roger and I were fairly skilled at endearing ourselves to the tenth-floor staff, but Al was a pro. Somehow he and Bernice helped rid the place of its awful strangeness. In the late afternoon they'd take their striding cardiovascular walks up and down the corridor, twenty lengths a mile, waving and greeting a multitude as they went.

When Roger had been on AZT a week, we were already buzzing to one another that it was working, as we watched him emerge from under water. The fear and labor went out of his breathing, and he talked merrily and tossed off puns. Even the looming IV equipment ceased to be so terrible, now that it was doing magic-bullet work. And despite the warnings of the doctor in charge of AZT that no one must know, since the drug wasn't licensed yet, I bragged to the nurses and anyone else who'd listen. One afternoon an IV nurse came in when Roger was dozing and demanded of me in a kind of spy whisper: "How did he get this drug?"

Her brother was an intern at Duke in North Carolina, and for several weeks he'd been telling her about the dramatic results, though he'd cautioned her that she probably wouldn't be seeing the drug at UCLA for at least half a year. Now she wanted to know everything about how we'd managed to spring it—a kindred spirit, clearly—so we took her through all the intricacies while she tapped the vein in Roger's arm. She pressed her beeper to tell them not to disturb her, she was in conference. This was the same woman who told a gathering of her colleagues that if they didn't like AIDS they'd better get out of medicine. She eventually came to work full time in the fourth-floor Immunology clinic. One of those remarkable people we'd meet occasionally along the way, who was passionately committed to facing the calamity, who wouldn't rest until she was part of the solution.

Jaimee flew in from Chicago for the weekend. We'd been reporting to her excitedly for days how much better Roger looked and felt, but of course she was stunned and upset by his gauntness. Yet the three days she was in town were wonderfully carefree. The two of them laughed their heads off, tracing the circuitous routes of their childhood: the trip to the Grand Canyon, where Jaimee was too young to ride a mule to the valley floor and fumed at the unfairness of it all; the visit to Disneyland in its opening summer, where they got to meet Walt himself on Main Street. "Like meeting Louis XIV at Versailles," as Rog used to say.

Late at night I'd take Jaimee back to the guesthouse at UCLA, where we'd sit over tea and unwind and strategize. Roger puffed with pride to see how seamlessly Jaimee and I got along. It gave him the greatest satisfaction, as if he'd struck a deeper chord in his family—for Jaimee was known as a guarded sort, highly selective about her loyalties. She remembers Roger getting out of bed to go across to the tenth-floor lounge, and fretting before he left the room, "Wait—is my hair brushed?" She couldn't believe it, the shy twinge of vanity about his bald spot, no matter that he looked half-starved. You prove you are still alive in the smallest gestures. Jaimee says she hadn't a clue how she looked herself that day. Like the rest of us, she threw on clothes in the morning and ran unkempt to Roger's room.

He could feel he was getting stronger again. At last it began to excite him as much as it did us, as if the future had opened again. "You're the miracle man," as Charlene the nurse would tell him one night the following summer after surgery. The passing of the crisis, the giving back of time—people with AIDS will tell you about that tidal shift, how it happens over and over. Now Roger would need all the positive feelings he could muster, because right after Jaimee left, the doctors didn't like what the arterial blood numbers were showing. They ordered another bronchoscopy, a test we had come to dread as a no-win situation. Once again Al and Bernice and I had to wait out the procedure in the lounge, and this time no one pretended to read.

They must have given Rog a jot too much local anesthetic, because he didn't come out of it as smoothly as before. For the next three hours he couldn't talk at all, just a whispered croak, and I had to put my ear very close to his lips to figure out what he was trying to say. At one point he was pleading for something, and I strained to listen and had him repeat, till I finally made it out: "My . . . balls."

They were jammed up under him because his hospital gown had got twisted beneath him while they were maneuvering him for the test. With the oxygen mask on and the IV in his arm, he couldn't move to release the pressure. His parents and I laughed with relief as I loosened the gown and gently pulled his balls free. Then about a half hour later he croaked painfully, word by slow word, "Why is this happening to me?" He meant the frozen larynx that made him unable to talk, but Bernice burst into tears at the larger resonance of the question. "I don't know, Rog," I said. "I don't know why anything anymore."

Yes, they told us briskly later that night, he did have PCP again and would have to go through yet another course of Pentamidine. I know we all had a terrible sinking spell at the news, and I recall

Roger gloomily announcing to people on the phone over the next few days: "I have three different diseases now." I was a wreck as I called around the country to find out just how bad a third bout of PCP could be, and Craig comforted me with the information that two people he knew had survived five rounds of it. As Peter Wolfe later explained, the PCP protozoa probably never went away entirely, and it wasn't yet understood what made it flare into full infection. The major advance they've made since then has to do with the prophylactic dose to keep the protozoa dormant. Many now receive a weekly dose of Pentamidine spray, inhaled directly into the lungs, which lowers the rate of repetition. But that wasn't available during our time in the war.

Since Roger would need to be in the hospital for another week, he asserted himself to get a change of venue, at least. One morning we moved him from 1020 to 1006, from a blue room to a peach room. I can still see that vinyl wallpaper in my sleep sometimes—the pulsating diamond pattern with a scroll border at the ceiling, blue or peach or beige. As a result of some benign bequest or other, Impressionist posters hung in each of the rooms. The Degas ballerinas that pirouetted all through the near-death experience had grown unbearably oppressive to Roger. In 1006 we settled for a placid Utrillo street scene.

By now, in spite of the constant stream of medications and the piggyback IV equipment, Roger was free for hours and getting antsy, so we'd go for strolls in the corridor. The other end of the floor, 10 West, was a world-class center for bone marrow transplants, whose patients all had complicated immunity problems and spent long periods in the hospital. We soon had nodding acquaintance with all of them and the masks they wore and their falling hair. In the evening we'd go down to the lounge, and I remember playing a game of checkers one night, a game neither of us had played in so long that we had to think hard to remember the rule of kings. Halfway through the game, though we were playing desultorily and nobody was winning, Roger pushed away from the table and said, "I don't want to play this anymore. It's too upsetting." I knew exactly what he meant. We were so drained and shaken still that even the automatic strategies of checkers were too demanding. We sat back talking till an old woman with a cane came in, accompanied by a doctor, and he quietly broke the news to her that her husband had Lou Gehrig's disease. We couldn't scramble out of there fast enough not to hear her shock and pain. The last thing she said as we fled was: "Doctor, we mustn't tell him."

As soon as Dennis Cope began to talk about a release date, Al and Bernice were insistent that we had to have some temporary help at home, "at least until Roger gets on his feet." We started asking around—nine to five, prepare two meals for Rog, light housekeeping—but there was no way we could hide the nature of the illness, even if we'd wanted to, which we didn't anymore. From here on, there would be no euphemisms on Kings Road. But we quickly discovered that a certain Geiger-counter effect had started, relative to AIDS, among the service professions. No, none of the temp agencies could fill the bill; perhaps we should try a full-time nurse.

Meanwhile Roger and I were worrying about money. He was keeping his office open and his secretary on salary, working as best he could by phone, his great tenacious goal to get back to his practice. Yet he wasn't even covering expenses anymore, and I was barely stealing an hour here and there with Alfred. We tried to assure the parents that we could do it on our own, though both of us remembered all too well the tension and craziness of the month before the darkness. Al and Bernice kept saying, "Paul, you've got to go back to work. You're the breadwinner now."

Bernice planned to stay on for a week or so after Roger came home, but she wanted Al to get back to Chicago. He was eating all wrong, his triglycerides were up, and he needed to see his doctor. There's a peculiar poignancy to the parallel trials of parent and child caught in the calamity. Al and Bernice never complained about their own health problems, though I knew how deeply ingrained was the worry in Bernice, who had monitored Al's angina for twenty years. In any case, they would be out again at the end of January, on their way to Palm Springs. We began to talk as if Roger would be back to work by then, and we'd be driving to the desert for a weekend, just as we always had.

An order was placed with the drug company for a two-week supply of AZT. For months we could never stock it any further ahead, as if they were making it in a kitchen lab with a two-burner output. We were told the drug would come in intravenous bottles, since a factory had not been retooled to produce a capsule. Apparently the drug had been on the shelf for years, awaiting the right disease, but it was a very expensive proposition to gear up to full market potential. The primacy of the market concern gives as good a picture as any of the chaos caused by the government in turning over drug research to private industry. But at least for the present we'd have our own piece of the rock. Roger would be drinking the drug directly out of IV bottles, three of them poured in juice six times a day.

I asked Charlie to pick up a timer one afternoon on his way to the hospital, and he arrived with a sleek digital item about the size of a package of cigarettes, with a tiny chamber for pills. I have six friends now on AZT—two white-banded blue capsules every four hours—and sometimes I'll be with one of them and the beeping alarm will go off, alerting them to the next fix. The sound of it always knocks the wind out of me, signaling such a confusion of hope and last chances.

And so we staggered home on December 11, with three shopping bags of drugs this time, and a warning that we must retain every last IV bottle and cap, to be returned when we came for the next batch. Burroughs Wellcome, the manufacturer, appeared to be quite paranoid about its trade secret. If we couldn't account for a bottle cap, who knew that we hadn't sold it surreptitiously to Upjohn? No question but that his homecoming was a blissful triumph for Rog, though there was still a shiver of disbelief in both of us, as if we didn't dare trust it yet. Meanwhile Bernice was indefatigable, appearing at our door at 7 A.M. so she could admit the nurse who came at eight for the final sequence of Pentamidine injections. Then Bernice would clean and cook all day, allowing me to huddle with Alfred at the computer in the study.

We had come to a dead end on every lead in the help department, and Bernice's dawn-to-dark labors bore mute witness to the need. Roger was feeling more himself every day, but it was equally clear it would be New Year's before he could get back to the office. Then we had a lucky break. Calling the Los Angeles AIDS Project to see if they could give us a referral on a nurse's aide who wouldn't be freaked by "the situation," we discovered there was a state pilot program just beginning, to study the cost-benefit of in-home care versus hospitalization. It made no sense to keep people in the hospital at a thousand dollars a day when they weren't even doctor-sick but simply debilitated or recuperating. The alternative was to maintain them in their own homes, providing daily help at a fraction of the cost. We offered to pay our own way to get into the study, but were told there were no financial requirements to qualify.

Next day we went down to the APLA offices on Santa Monica Boulevard, across from Plummer Park. Not an easy place to enter. I remember one of the founders of GMHC in New York telling Craig how he'd hate to need any of the services he'd created, not because it was demeaning to ask for help but because the issues raised were so awful—lost insurance, lost jobs, evictions, the full gamut of miseries. Roger and I had spent years blithely writing checks to such organizations, and surely there is magic in that as well. One does it in part to cover one's ass, knocking on wood: *Please, not me.*

Now we sat on a battered sofa, staring across at a safe-sex poster, while I brooded about the germs on the grimy upholstery. Two men were cheerfully answering phones, coaxing people to ask the questions that had frozen in their throats. Then a counselor took us to an inner office for an intake interview. Roger answered all his questions simply and without self-consciousness. I was the one with the hurt pride. It was one of those irrational moments when I wanted to cry "Time out" and trot out Roger's degrees and a list of my credits, to protect us from the ignominy of it all.

Thankfully, we qualified for the program. They would try to assign someone as early as Monday, the day before Bernice was leaving. We walked out of there relieved, our dignity remarkably intact. Through the trials of the next ten months our dealings with the Project were notably life-affirming, with a hands-on human touch that never wavered. When we got home, Roger called Sheldon to tell him we'd found someone, but when he told him where, Sheldon clucked with disgust and said, "Somebody's got to put a stop to this." Implying that we were stealing from the indigent, offended that we should be stooping to charity.

I was so sick of people's opinions. Unsolicited advice comes pouring in from those who can't be really there, till you feel like a laboratory of other people's whims. We were having a hard enough time believing our life was still our own. Later that night, on our walk, I gnashed my teeth with anxiety. "What's happening to us?" I asked in desperation. "What are we going to do?"

"Paul, we have to accept our fate," said Rog, firm and unsentimental. "There's no other choice."

"But I can't," I whined, and I meant *I won't*. Yet even as I said it, it struck me how Greek Roger's attitude was. I can't express how small I felt just then, or how alone, as I looked at Rog in the dark and understood he had reached a kind of acceptance. I still wanted Greece to be sunny and exalted, with white stone ruins and statues of gods so perfectly human they breathed. Beauty was as far as I needed to go, and I wasn't equipped for the tragic design of fate.

But being home was so seductive: Within days we'd started to make plans for Christmas. Craig and I had been leaving it open for weeks now as to whether he'd come for the holiday, and in fact he didn't decide for sure till the twenty-third. Now I began to coax him to make the trip, and Roger encouraged me to buy a tree and have people over on Christmas Eve, as we always did. We would keep the whole thing on a much smaller scale, but I think he figured it as a way for me to proclaim we were still in the game. He himself was eagerly making plans to be back to work right after Christmas, and I told Alfred we could go ahead with the horror script for Warner Brothers. All this AZT optimism was better than a Currier & Ives snowfall.

I knew, of course, that Bruce had been gravely ill during all the weeks Roger was in the hospital. I even managed to check in every few days with Chana, in the hope that I could at least talk to Bruce and let him know we had copped the AZT. I realized only later how carefully his friends were couching what they said to me. I had some indication that there were episodes when his mind wasn't right, but dementia was still largely undefined and unspoken, even in the AIDS underground. I heard that nurses had been brought in around the clock. But none of it struck me as fatal: It had been only a few weeks since I'd seen Bruce myself, and it was Roger, after all, who had just nearly died.

Bruce wasn't even hospitalized, so I kept assuming he must be treading water, the way Roger had before his crisis. Surely one day soon I'd call and he'd actually answer the phone. But it was always Chana who answered, and she would say carefully: "Bruce's case is completely different from Roger's. He's been a lot sicker from the very beginning." She was trying to break it to me easy that he wasn't going to make it, but the subtlety went right by me. They obviously didn't understand how close as a shadow death could be and still you could squeak through and outwit it. In any case, it was only a matter of weeks before we'd force them to expand the AZT protocol. Bruce just had to hold on.

I woke with a start on Sunday morning the fifteenth when the phone rang. I was sleeping in late again because Bernice was covering the morning shift, and this was the day I meant to go get the Christmas tree. But I panicked at the sound of the phone and scrambled out of bed, instinctively feeling I had to get it before Roger did. Then I heard him say hello from the back bedroom, and as I came out into the hall he gave out a low wail of pain: "Oh, no." I stood dumbly in the bedroom doorway as he looked at me in total defeat. "Bruce died last night," he said.

I felt the same spurt of annoyance as when Ted called a month before to tell me Cesar was dead: Why are they bothering Roger? I had no time to mourn Bruce either. At one point I wanted to go over and see his sister Carol, who flew in from New York, especially when I heard she was asking for me. But I was afraid to walk the two blocks and be in the presence of death, afraid I might bring it back with me or see too much of the apparatus of mourning. If there had been a funeral I would have gone, but they decided to put off the memorial till mid-January.

Six months later Chana called to tell me Bruce had left me the huge Francis Bacon lithograph in his will, because it was a picture of a writer. And I thought: He left me that because I got him the drug that killed him. By then Roger was blind, and I wouldn't bring a new picture into the house if he couldn't see it, so the Bacon never arrived till six weeks after Roger died. I still haven't hung it up.

It all came full circle in August '87. The final flourish in Bruce's will was the wish to have his ashes scattered at Fire Island, where he'd played out so much of the glamour of his youth. The *most* beautiful place, as he would have said, with the *most* beautiful boys. I flew east to join the family at the ferry slip in Sayville—his parents, his sister, his friend Jimmy—and we went over on a milky summer morning, the plan being to toss the ashes from the ferry window into the bay. Illegal, of course: ashes belong in the open sea, beyond the three-mile limit. Bruce's cousin passed the box around, and I watched Jimmy cradle it and start to cry. Then he passed it to me. It was heavier than I expected—the box was bronze—and it felt truly as if

I were holding the final weight of a man. That's when I cried for Bruce. An hour later on the dunes I cried for Cesar, whose ashes I never held, dispersed I know not where. Then I cried all the way back to L.A. on the plane, for Roger mostly by then, but really for all of us, this generation of widows and groping survivors.

The first attendant we had from APLA was Jack, a bald, enthusiastic fellow who bustled about the house dispensing cheer as his marinara sauce simmered on the stove. It was such an odd time for him to be there, because I was in the middle of Christmas preparations, readying the loft for Craig, and Alfred and I were brawling in the study as we brought a script to completion. The house on Kings Road didn't feel like a hospice at all, and Roger seemed very much part of the bustle, though I could see Jack attune himself to Roger's pace, to the rhythm of his naps—quietly tidying cupboards and reading Theosophy during his breaks. I felt guilty eating anything he made, since he was supposed to be cooking for Roger. Yet Jack didn't seem to mind the upbeat air in the house at all. He'd already been through a couple of grim final stages, and plainly we were a picnic by comparison.

About three days before Christmas, Roger got word from his secretary that she would be leaving in February and moving back to the Midwest. Ricki was the best assistant he ever had, and she'd covered all his bases at the office with superhuman skill for months, never breaching the discretion of the unnamed sickness. Her giving notice was a blow to Roger, who thought of her as his last vital link to Century City, the office he hadn't been in for six weeks. I swore we'd find somebody else as good, but that night when Roger told his brother on the phone, Sheldon said, "Maybe this is a blessing in disguise." He clearly felt it was time to close the office, and I can't remember seeing Roger as depressed as he was after that call.

But that same night he also took a call from a friend in Boston, an obsessive woman who'd fallen in love yet again with the wrong man and whose mind was racing with self-deception. Roger listened with understanding and compassion, and then bluntly confronted her with her acting out of the same old pattern. She still speaks of his extraordinary clarity that night—how he was able to shake free of his own dilemma, determined that nobody else waste any more time. This dynamic would repeat itself over and over through his illness—friends would call with clumsy words of comfort, only to find themselves opening up to Rog and hearing him comfort them.

"But that's not important," I'd hear him say bemusedly, though never by way of dismissal. He'd tease his friends about their timeworn frets and quibbles, then heap them back on himself with a laugh. The phone was becoming a lifeline to his past as people caught up with the secret, and the one-on-one conversations he loved best never failed to lift his spirits. There were times when the phone link was the last thing he was left with, aside from me.

Once more the box of decorations came down from the attic, and Roger, dozing on the sofa in the very position Cesar had assumed the previous two years, watched me put up the tree. I was too happy to have him home to dwell very much on the sad losses of the year in between. Richard Ide came over on Saturday the twenty-first, before leaving town for the holidays; he had an armload of packages, Roger having instructed him what to buy me for Christmas. Richard cooked lunch and announced as he served it to us on the terrace: "The secret of Sloppy Joes is the buns." Roger and I cracked each other up repeating that line for months, or sometimes we'd murmur it perfectly straight-faced while sitting in endless waiting rooms.

Those first few weeks of AZT, I didn't trust Roger to wake up in the middle of the night, or perhaps I just needed to be around for every dose. In any case, I bolted awake at two and six when I heard the beeper. I uncapped the bottles, poured the elixir into the juice and gave it to Rog, who happily stayed half asleep through the whole procedure. Every dose was a miracle to me. On the twenty-third I ran over to UCLA for the next batch, taking with me a bagful of empties. There was something chaotic going on with the protocol, so Suzette had to officially *not* give me the next week's supply under the table, while she untangled the red tape. As I was leaving, the doctor in charge came up and asked how Roger was doing, and I waxed eloquent about the drug, saying that everyone ought to be on it. He seemed worried that my enthusiasm would find its way into print. Several of the doctors, burned the previous summer by the Rock Hudson media show, were skittish that I was a writer.

Craig arrived the night of the twenty-third, and we went into immediate overdrive to prepare for fifteen guests on Christmas Eve. Craig recalls that Roger was in and out of the bedroom the whole week of his visit, up for a couple of hours, then down again for a nap. He seemed more fragile than frail, and of course he carried an aura of marvelous expectancy wherever he went, on account of the AZT. I realize now that a great change had set in after the long battle of autumn. Between mid-October and mid-December Roger had crossed a minefield, and the price he paid to get through to the other side was that now he would need to rest an hour for an hour of strength. But that was a bargain we were both so grateful to strike after the pass with death that we hardly noticed it at first. The cup was half full, period. If it took twelve or fifteen hours of rest and sleep every day to process coming back to life, that only made the remaining time more festive. The minor presents we wrapped were props and tokens of the occasion, but the real gift—we all knew this—was being consumed at a rate of eighteen bottles a day in the back bedroom.

Just before Craig flew to California, he'd banded together with four other men with AIDS to form a group very like a platoon. They'd all been dissatisfied by the groups they'd sat in on—too much negativity, whining and bitching about hospital crap, the blank-stare bureaucrats at every turn. Craig's group wanted to get out and find the magic bullet. None of their doctors seemed to agree about alternative therapies and the state of antiviral research. Since each had high contacts in one establishment or another, they figured they'd come together to share the scuttlebutt, the rumors and leaks. The group included Vito Russo, the incandescent film historian, all Brooklyn edges and a foot-wide grin. Since AZT was at the top of the list of what they wanted to know, part of Craig's visit was in the nature of a field trip.

"You're the AZT poster child," I'd tell Rog as he drank his dose.

The party on Christmas Eve was strange but not without cheer. Alfred and his friend Barry Miller arrived directly from Tom Kiwan's funeral. Roger slept in until everyone had assembled, then made an appearance for an hour or so, talking quietly with one and another as they slipped into a chair beside him. Almost everyone there had been coming to our house on Christmas Eve for years, and I expect that the great change between '84 and '85 was more difficult for them than for us. I don't think anyone fully understood how close we'd come to the edge in November or how charged with hope we were again. After a while Roger excused himself to lie down again, and most of the guests took their leave of him from the bedroom doorway.

What did we do that week? I've asked Craig since, and he can't remember us going anywhere. In one way it was intrinsic to the season that people came to us instead. Robbert and Susan dropped over on Christmas Day, and their gift to Rog was a photograph from Robbert's *Midwest Diary*, a plain of new-fallen snow, with telephone poles and a stop sign. It embodied the spare poetry Roger loved in art, from the wire circus of Calder to the cloister at St. Trophime. I think I must have had a certain need to stay close to the house, now that we'd finally been given the house back. I knew how dead and empty it became when Roger was in the hospital. There was also the matter of the four-hour rhythm and never wanting to miss the dispensing of AZT.

Mostly Craig and I just talked about being alive, and I swung back and forth through the house like a yo-yo, from living room to back bedroom, checking up on Rog. By now there was a fairly intense political consciousness evolving in New York, and we in L.A. were as far behind that as we'd been behind the disease in the first four years of the calamity. The indifference of the press remained deafening; AIDS activists liked to talk about the occasion when the *New York Times* devoted front-page space to a disease that felled seventeen Lippizaner stallions in Europe, when no story about AIDS had ever appeared on page one. The morass in Washington went on and on. And yet Craig and I insisted we'd keep fighting, become spies and outlaws if we had to, but we wouldn't go quietly.

I'd had the same caution as Craig's platoon about the focus of many AIDS programs, which seemed designed to help people write their wills and memorial services. The old medical chestnut—*Hope for the best, prepare for the worst*—had been truncated of its better half. Craig and I decided we didn't want to be in touch with letting go, not by a long shot. Yet we bristled, from the other side, at the growing "empowerment" movement, which tended to start with the assumption that people brought on their own illnesses. From there they moved to the notion that once you could fully love yourself the virus would evaporate like fog. The whole guilt-and-redemption trip was much too Catholic for my taste, besides which it seemed to consign those who died to the status of losers. Also, several of the self-healers were against any form of medication, while we were engaged in a major struggle for antivirals. Happily, these positions have become less rigid in the last year or so, and there is considerably more cooperation as people with AIDS dictate the agenda of the fight themselves.

But what I remember most about Christmas week with Craig was telling him how much I loved him. Loss teaches you very fast what cannot go without saying. The course of our lives had paralleled the course of the movement itself since Stonewall, and now our bitterness about the indifference of the system made us feel keenly how tenuous our history was. Everything we had been together—brothers and friends beyond anything the suffocating years in the closet could dream of—might yet be wiped away. If we all died and all our books were burned, then a hundred years from now no one would ever know. So we figured we had to know and name it ourselves, tell each other what we had become in coming out. I also believed that Craig understood what Roger had been through, as much as anyone could outside us. Every night the three of us gathered for dinner, lazy and content with Christmas leftovers as we recalled the snows of Beacon Hill. When Craig left on the twenty-ninth, the house was suddenly terribly quiet, in spite of Jack's post-Christmas hum of positive thinking.

My friend Star was planning to come through L.A. from Hawaii over New Year's, so I resisted the impulse to haul the tree out to the trash with all its ornaments. I certainly didn't relish boxing them up again, contemplating all the while who might be gone by next year. Then suddenly, on the thirtieth, we had a call from Dennis Cope, saying Roger's white blood count had fallen precipitously over the previous few days. He was to go off AZT immediately and come into the hospital for tests. It was a cruel upheaval—he'd been home only seventeen days—and we packed in stunned and wounded silence.

We didn't really understand that bone marrow suppression was a typical drug reaction. In any case, it required two days of renewed terror before they could establish for certain there was nothing else going on. Now we would learn the subtle enslavement of the numbers. The white blood count was healthy above, say, 3,000, but after a few weeks of AZT it would start to swing down. If they didn't halt the drug the count would fall below a thousand, leaving Roger at risk for all manner of bacterial infections. It tended to hover stubbornly at its lowest ebb for several days, then creep up again as we waited with mad impatience for the restoration of the magic bullet. When the count was very low, Roger had to be in the hospital under protective isolation. It was two weeks before he got home again.

But it took days for all of that to make sense, and for us to believe the count was really going up. I remember sitting on his bed on New

Year's Eve and holding hands—the hospital unbelievably quiet, everyone who could still crawl sent home for the holidays—wondering if the roller coaster would ever stop, and resolving on the stroke of midnight that nothing mattered as long as we stayed together. We had finally had our introduction to the elaborate procedures of isolation. For the first time I had to put on a blue mask, like the ones the more jittery nurses had been donning for some time. Now everyone had to wear one for Roger's protection, and gloves and gowns in addition for the hospital staff. Cope told me I could do without the gloves as long as I washed my hands thoroughly. It took me only a matter of hours to reorient myself to the new crisis, keeping germs away. I washed my hands about every fifteen minutes, a habit I can't seem to shake even now.

Star, who visited on New Year's Day, recalls that Roger looked startlingly well, his weight up again and his color good. But I was a total wreck, and I railed at her for waiting a day to come by, for cornering Gottlieb in the hall to ask him if she was at risk, and she finally took me on a walk through the UCLA grounds and let me talk out the madness. I begged her to come to the house next day to help me take down the Christmas tree, and when she did she saw for the first time how transformed the place was, with its rows of medications and cartons of AZT. Before she left L.A. she wrote out a three-page list of daily things to do, an attempt to get me to focus and stop careening off the graph. I would stare at the list every morning—PRESENT ROGER WITH AN UPBEAT ATTITUDE, WHEN EMOTIONAL DO NOT ACT RIGHT AWAY, GET EXERCISE—and I'd stay on an even keel for a couple of hours, then collapse into panic again.

The big decision that couldn't be put off any longer was the closing of the office, and Sheldon was probably right to keep steering the conversation that way. The situation had grown too unpredictable: Roger couldn't run an office he couldn't get to. Sheldon felt it was time for him to pull back and devote his full energy to regaining his health. Within a few days Al was being similarly supportive, assuring

Rog it would be no sweat to reopen the office once he was on his feet—that phrase again. None of it went down easily with Roger, who saw the loss of his work as the beginning of the end. I remember him turning to me and saying in a helpless voice, fighting back tears, "But what about all my files?"

That one line is as painful to recall as anything not strictly medical during the whole course of the calamity. His files were the accumulation of his years in California, both his lawyering and all our common interests. There's an enormous amount of basic research and format material you have to have to service a private practice, and Roger was proud of the range of his files and their rigorous organization under Ricki. I kept thinking of the photographs on the office walls, so carefully chosen to startle but not intimidate, and the black leather Italian desk chair I'd given him when he opened for business.

The most coherent decision we made during the crisis over the office was that he bring all the files home and work from there. It was a bit unsettling to contemplate how this arrangement would work in practice, in a house where there was already one ragged sole proprietorship in operation. I couldn't imagine where everything would go, but as we talked about it more, Roger grew confident and ready to give it a try. At least he would have Ricki for a few more weeks, to separate files and pack up boxes. Meanwhile he was working in the hospital again, difficult though the logistics were. His drive and energy had returned, a more significant measure of his status as a lawyer than where he hung his shingle. Cleaning out drawers, I still come upon pages of yellow legal paper, full of notes on contracts and partnerships. Though they sound like Martian to me, if they're dated January '86 I read them through amazed, till I want to cheer at the sheer concreteness of them.

Once we understood that Roger wasn't actually *sick* with anything, we went back to singing the praises of AZT and how well it was making him feel. But things were veering into chaos again between the two of us. Roger's feeling stronger only served to accentuate his impatience at being holed up on the tenth floor, and I was trapped in a deadline with Alfred and squabbling every time we sat down to write. It's not that Roger and I were bickering, exactly. But I was overreacting again to every slight and turmoil, and he couldn't take it. A couple of times he bellowed at me to get out of his room, dissolving me in tears. Sam had been pushing for weeks now, ever since Roger came out of the woods, that he needed some kind of counseling. Fortuitously, a psychiatrist we knew—Ronald Martin—mentioned to Rand that he'd like to help if he could. The protective hospitalization finally tipped the balance for Rog, and he decided to see Dr. Martin once a week following his release.

Roger came home the eleventh. On the fifteenth I went to Bruce's memorial service. I felt calm driving over to Bruce's house, because I'd left Roger out on the terrace chatting happily with a friend. The service was on the bluff in front of Bruce's house, and I sat at the end of a row and stared out over Hollywood while several people made quiet remarks—including Marisa Berenson, the sight of whom made me think I must call Cesar, who'd always loved her style. Pat Ast arrived late in a cab that couldn't make it up the hill through the chaos of parked cars, and she shrieked as she ran up the steep driveway: "Wait! Wait!" She came barreling in in her full *zaftig* eminence and read a childlike poem to Bruce. All through the service I just kept thinking, *Rog, please don't die.*

A couple of days later I had a meeting with my Hollywood agent to discuss the splintered progress of my career, and he said categorically, "You can't work anymore, not with your lover sick like this." He probably meant to suggest that I was too busy being a nurse, but it came out sounding as if he didn't want all this unpleasantness around him. Then he wanted to gossip about Stan Kamen, the superagent at the William Morris Agency, who'd just informed his stable of megabuck clients that he was taking a leave of absence on account of liver cancer. Kamen's was the most popular case of AIDS in Hollywood at the time, a source of endless gossip for the barracudas.

Then the agent asked if Roger was "covered with sores," and I informed him icily that lesions were not sores, and no, he wasn't. But I swallowed all of it and begged him to track me down a novelization for the summer, since I had a bit of topspin in that department and thought the work might not be too taxing. Oh, yes, he said, leave it all up to him, but he never called me back.

Ricki came by one afternoon with boxes of client files, so Roger could begin to regroup his active practice at the house. And we began the mostly benign fencing match as to whose turn it was on the word processor and when do you think you'll be finished with the phone. Yet it was clear we were going to make the details work, and once again we played it as if the future would be free of further complication.

My happiest memory of the window of mid-January was a dusty Saturday afternoon downtown. We'd gone to UCLA for a set of blood tests—they drew it every third day now, trying to keep ahead of the white count—and afterwards drove in to USC, deep in the old city. We wandered across the campus, poking in courtyards and admiring old trees, ending up at the university museum, where Susan Rankaitis was having a show. Two pieces from the Icarus series were on exhibit, including a huge diptych called *The Rog Chant*. Its centrifugal explosion of abstract forms was intercut with the repeated image of a figure with arms crossed over its face, as if blinded by the burning of the sun. The show had been up for several weeks, and we were the only ones there for a while. I remember Roger standing in front of it, studying it with utter concentration, as if he owed this one the deepest look he had. It was the last time we were ever in a museum together.

The window lasted about two weeks, and again we had the sudden call ordering Roger off AZT and back into the hospital. This time the white count had fallen to 500, and this time it really felt like the last straw. "I don't want to spend the rest of my life in the hospital," Roger declared with furious anguish as we drove over once again—the overnight bag packed automatically now with slippers and robe and toilet kit, a couple of changes of underwear. I happened to be strong at that moment, spurring him forward, making him see that he'd recovered beyond our imagining in the six weeks of AZT.

Sam tried to sort out the feelings behind my rage at minor things and Roger's corresponding sullenness and withdrawal. Somehow I had to get it across to Rog that I was angry at the illness, not at him. Our lives had utterly changed, and though I was willing to accept those changes, I hated them still. Meanwhile Roger was doubtless feeling guilty about the ways in which his illness had shattered my life. My falling apart only deepened his sense of having hurt me. On January 30 the Jaguar—possessed by a devil, clearly—locked in gear again in the parking lot. I proceeded to go bananas, an instant replay of four months earlier. Roger screamed at me to call Triple A and leave him out of it. The car was merely a triggering device, Sam said, symbol of how meaninglessly things kept going wrong. But I simply had to control my lunacy over these everyday problems: Roger needed the positive aspects of normalcy, not the mess of things gone haywire.

Unfortunately, our scrapes and dislocations were closing off communication instead of opening it up. We needed crisis management, needed to devise a better radar for giving one another what we needed. Sometimes, after all, I'd be rattled or weepy, and Roger would be right there to prop me up. "I know you so well," he'd say tenderly, tousling my hair and forcing me to smile. "I know everything about you." But if he was having a day that was especially bad and he wasn't coping well, I needed to know that. Otherwise we were stuck with this clash of my volatility and his closing off.

I had an appointment with Dr. Martin myself, in an office right out of a *New Yorker* cartoon, with its analytic couch and its Freudian wall of books. I told him how upset it made me when I saw Roger in a sort of infantilized state with the nurses, relying on them too much. I admitted I wasn't doing well handling Roger's particular brand of mood swing, from hopelessness to a sort of nonchalance, while my own emotions suddenly seemed inappropriately overscale. I realized it was hard for Rog to understand, assaulted as he was, that there were two damaged people here. In this perpetual state of tension, this waiting for the other shoe to drop, even normal wasn't normal anymore. How could we plan on anything? How could I go back to work with all this rage and fear? I remember Martin talking intelligently and helpfully, but I took only one note in his office: *I can't tell Roger not to die. It's not fair.*

Somehow we pulled together and talked the deadlock through, and it never got quite so out of hand again. Meanwhile we groped for ways to alleviate the imprisonment. Since Roger continued to feel energized, we took long walks in the courtyard outside the hospital, and a couple of times even ventured through the botanical garden that filled the shallow canyon beyond the emergency entrance. We managed to cop some afternoon passes so Roger could come to the house for lunch, to sit in the garden and play with the dog. As we drove back and forth in the car, one of us would always wear a mask, and people stared at us in traffic. After Roger died I found pictures on a forgotten roll of film in the camera: He's holding his blue mask in one hand, the tree ferns fanning behind him. Now I see how changed he was since the previous pictures, from Massachusetts in August. He's balder and more frazzled, and he's got a belly from all the feeding, but he manages the gentlest smile, and the imp is still in his eye.

Dr. Martin said we should reminisce more, so I hauled all the Greece pictures up there, even all the guidebooks, and we talked our way through the Aegean again. One day when Rog was especially stir-crazy, I brought a new address book with me, and he spent a couple of afternoons transcribing addresses from the old—in sensible pencil, of course, so changes could always be made. Over Christmas

I'd started to read through an old pile of *National Geographics*, the magazine that is designed to end up in a pile, and I'd report to Roger on all these exotic places. By way of armchair travel we especially liked to talk about digs, since they inevitably reminded us of sifting through the palace ruins in Crete. One Sunday I ducked out to Westwood Village and bought a basic archaeology text, and for the next several weeks we covered Egypt, Troy and Mesopotamia, pricking memories of Roger's year in the Middle East.

These were the days of February when the Marcoses were toppling in Manila, and there was considerable flurry on the tenth floor as the news broke. Two of our favorite nurses—Rosemary and Luzminda—were Filipino, and the drama being played out was their own. I realize now how the headlines served to pull us together up there, on a sort of neutral emotional ground. We were there the day of the Challenger disaster too, and later on in the spring for the strafing of Libya. It was a sign of engagement to be able to pay attention to any sort of crisis beyond the four walls.

One morning the L.A. *Times* had a long piece about Roy Cohn and the rumor—rigorously denied—that he had AIDS. One of the AZT doctors was in, and we talked about the futility of stonewalling but also tried to figure out the psychology of it, since one man's secret was never quite the same as another's. The doctor let slip a remark about a movie star, not naming him by name, but I knew right away who it was. We in the underground had been fielding speculation on the case for months. There had been a recent sighting of the star in question at a hospital in Houston. Craig and I had decided that two sightings were required to make a rumor stick, and the doctor's implication that he'd treated X at UCLA sewed up the case. This didn't mean we necessarily *spread* the rumor, but we got a certain satisfaction out of running it to ground. Roger didn't care about any of this, not the way Craig and I did. Somewhere in that knot of our anger and despair there was a *National Enquirer* streak that wanted celebrity blood. Because part of what allowed people to keep the disease at a distance was believing it couldn't touch anyone they cared about.

When did the shingles appear? I didn't pay that much attention, frankly, not at first. One day there were a couple of spots on Roger's stomach, and we immediately paged the intern, fearful it was some kind of drug reaction, or even worse, KS. When they said it was shingles, all I could think was that it couldn't kill him. Half the gay men I knew had had shingles. But three or four days later it was a flaming crimson sash of blisters across the right side of his rib cage, and it stung on that maddening knife edge between itch and pain. There was nothing to do to relieve it, and it went on for weeks, the pain echoing long after the rash had faded.

The shingles must have started during the late-January imprisonment, then, not the one over New Year's, when the white count was all we could think of. There are two cheerful entries in my journal for February, so he can't have been suffering the worst of the shingles still. Though I can't pin down the date exactly, I can see him wincing from the pain, arching his back and barely holding off from clawing the blisters open. I'd apply cold cloths and, when the itch became insane, paint him with great swaths of calamine lotion.

I remember one especially bad day of shingles in the hospital, when we had a visit from a brainless couple who said all the wrong things and brought up gloomy details of the sudden demise of others, till I thought one of us would scream if they didn't leave. But as it happened, they segued into real estate, and talked of a house for sale on their street. Roger perked up instantly and grilled them about the particulars. Richard Ide had been looking for a house for months, and he and Roger talked endlessly about financing and neighborhoods. Roger couldn't wait to tell him about the house on Holly Drive, since it hadn't gone on the market yet. Within a matter of days Richard had made an offer on it, and all through escrow Roger reviewed the paperwork. It made him feel terrific that he could still accomplish something, even from his tenth-floor cell.

When he finally got in to see Richard's house he was practically blind, though all through the final summer we would have happy times there. And as we'd drive away to go home, Roger would always say, "Isn't it great that he found this place?" Not quite taking the credit.

He was home again at the beginning of February and back on elixir. The shingles clung like barnacles, but we never really thought of them as an opportunistic infection, though of course they were. We hadn't really thought through the scary notion that AZT had somehow let an infection in. All we cared about was that he was on it again, and we felt the roller coaster click into its uphill rise and went with the ride. Sunday, February 9:

> *. . . today we were up at Will Rogers* [*State Park*] *in this aching sunny weather, & we ate on the lawn & then took a walk & watched 3 horses crop a hill pasture & we sat on a log & had this nodding acquaintance with what a moment is.*

"Wouldn't you like right now to last forever?" Roger said as we lolled on the picnic slope. The afterglow of it lasted for days, for we bragged about it as if we'd just ascended K-2.

Then Roger's aunt and uncle arrived from Jerusalem and put up at Sheldon's for a couple of weeks before going on to Palm Springs. Rita and Aharon would visit with us every day, smoking like chimneys and regaling us with all the contortions of Mideast politics. Aharon Remez, whose father had signed the Israeli constitution, had been Israeli ambassador to Britain in the late sixties, then head of the Weizmann Institute, and most recently director of ports. Rita was Bernice's sister, passionate and intuitive, heart on her slightly unraveled sleeve. Their visit happened to coincide with a swath of stable time, and Roger was flush with delight to have them around. They were utterly unpretentious, and our sort of bohemians. Rita arrived laden with material about imaging, a process we'd barely heard about. She took Roger through a lot of exercises designed to isolate and control the shingles pain, and even I managed to sit down long enough to learn a few techniques.

We made a plan to meet Rita and Aharon at Sheldon's for lunch on the Saturday we dropped by Roger's office to check on the packing. As we came into the empty lobby, signed in and rode to the twenty-fourth floor, I thought of the countless times I'd gone with Rog to the office on a weekend, taking with me something to work on quietly by his desk, so neither of us would have to be alone. I don't know if we bothered to try to prepare ourselves for the shock, but nothing is like reality after all.

We got to the door that said *Roger D. Horwitz*—a sign that's still propped among the boxes in the garage, because I can't bear to throw it away—and swung it open. In the middle of the floor was a pyramid of boxes. The photographs were stacked against one wall, the empty drawers of the desk pulled out as if they'd been ransacked. "Oh, God," Roger said, covering his face with his hands and starting to cry.

I couldn't think of anything to say, but the line that ran through my head was, "This is the saddest story I have ever heard." It's the first sentence of *The Good Soldier*, which Roger was given to quoting admiringly; it had hooked him like the beginning of no other book he'd ever read. Now I think about that moment in the office as a point of no return, as the moment in which he saw in a single look how very much he'd lost.

Yet he was stronger and stronger—at least we said he was—and we started having people in again for dinner, though we were too scared of the numbers to venture out to a restaurant. The doctors tried to tell us we were getting too fixed on the white count, but how could we not, with the magic bullet dependent on it? Even when the count was up and Roger was beeping every four hours, we shied away from crowds. Saturday the fifteenth we went over to the Perloffs' for dinner, along with Susan and Robbert. Roger wore a mask except when we ate, and I insisted we set up a tray table for him six feet away from the dining table, an elaborate and probably absurd precaution. But by then we just had to go with our instincts, and our friends did what they could to make us less afraid. That night the Perloffs showed their slides of France and Italy—circa 1955—for Susan was about to go to Europe for the first time. We laughed at how young the Perloffs were and chattered amiably about various totemic sights, but I burned with longing and jealousy to think we would never go back, never see Paris or Madeleine again.

The office move was scheduled for February 25, but Rog had enough of his files at home now that he was working along smoothly, picking up business. We understood there might yet be a revolving door into the hospital until they'd figured out the proper dose of AZT, but meanwhile we had days and days in which we simply got to live again—however battered, and Roger constantly fighting the pain of the shingles, but going with the good times.

I think we must have been very low on our stock of AZT, for though the company had finally retooled itself to dispense it in capsules rather than bottles, there was still a lot of suspense as to when the next batch would come in. I was scheduled to make a run to UCLA for a pickup, and Al and Bernice, in town from Palm Springs, were on their way over to visit with Rog while I was out. He had finished lunch and was in the living room working. I was in the study, and I called out a question to him, but he didn't answer. When I went in he was sitting quietly on the sofa. I asked him again, though I can't remember anymore what the question was.

He began to speak, halting and slow: "Um . . . I . . . um . . ." His face was perfectly calm, and there was no panic in his voice. It was as if he was fishing for a word.

"You what?" I asked. More hemming and hawing, as if his mind were on something else. *Absent*minded. I was frustrated rather than scared. "What are you trying to say, Rog?"

With agonizing concentration, struggling with every sound, he spoke: "I . . . um . . . have to get . . . um . . . my medicine."

Now I could feel my knees turn to water. "What's wrong?" I

gasped, unnerved by the clench of his voice and the eerily quiet tone. He didn't answer. I wanted to shake him. "Tell me what's wrong."

"Shut up," he cried, and then he seemed to sink into himself. Now I could see fear in his eyes.

"It's okay, Rog, it's okay." I sat on the coffee table facing him and held his hand. "Are you all right?" He nodded, but he couldn't talk. It was as if a curtain had been drawn across the speech center in his mind, and the only thing urgent enough to pierce it had been the need for more medicine. As if he understood instinctively how desperate the moment was, and his last plea was for AZT. I wasn't sure how to keep him calm and get him to the hospital. I wasn't sure what I would say to the parents, who were due in five minutes. I squeezed his hand; he squeezed back. All I knew was that we were in terrible trouble.

After a few minutes he was able to answer me "yes" and "no," but when I stupidly tried to return to the original question he started looking panicked again. We went out for a walk up Harold Way, and I said we'd go over to UCLA together while I picked up the AZT. We'd call ahead and alert Cope so we could tell him about this episode. Roger was clearly very nervous but nodded agreement to the plan. When Al and Bernice pulled up, I informed them casually enough—as Roger hung back tense and shy—that he'd had a momentary blank, and they should follow us to the hospital. I recall remaining very sane all the way over: I kept asking Roger simple questions, so as to hear that reassuring yes and no.

By the time Cope came into the clinic examining room, Roger and I were talking normally, if a bit subdued. There was a fragile air about Rog, worry under the shell shock, as if he'd just come through a train wreck without a scratch. Now as he sat in a straight chair and Cope prodded him with questions, he kept rubbing his hands along his thighs in a nervous tic. He answered every question clearly, but his voice was thin with emotional overload. Cope's instinct was that the whole episode had been an anxiety attack, triggered by an irrational fear that there might be a break in delivery of AZT. To be safe, however, they'd better admit him for a couple of days and have a neurologist examine him.

While he made phone calls to set that up, Suzette came down from Immunology with our next cache of the drug. As she and I were making the exchange in the hall, I tumbled out my relief that all we were here for today was stress—though the unspoken fear of viral invasion of the brain had left me in a state of shock as well. Suzette was gentle and supportive as always, but she made some reference to how we would handle things' taking a bad turn. "Of course the two of you have talked about dying," she said.

"No, we haven't," I answered with some defiance. "We talk about fighting."

She gave it another shot, agreeing that the best course was hope, but even so we had to be realistic. The long-run chances were slim. But it went right over my head, so utterly would I not countenance any talk of death. Fortunately, the next two days' hospitalization bore out the notion of false alarm. Roger ran another gauntlet of spinal tap and bone marrow biopsy, both clear. There was a long session with a neurologist, who asked him to repeat a series of numbers, then showed him pictures of *People* types, all of whom Roger identified except the football player. How many fingers; eleven times six; who is Cher? He came through swimmingly, though the whole time I was as nervous as a mother at a spelling bee, telegraphing answers, heart in my mouth if he paused for a split second.

The neurologist concluded there was nothing manifestly wrong with the nervous system. I was so relieved I could have kissed him, though in fact he was a chill sort, who didn't cotton one bit to my encyclopedic take on AIDS. He would not accept me as part of the situation. He was also an example of a curious phenomenon, the doctor with matinee-idol looks, about which my brother and I used to theorize at length. Bob had seen more beige hospital walls than I had, and we agreed about that certain species of good-looking specialist. Our theory was that pretty people got spoiled and coddled all their lives, and though their looks alone presumably didn't get them through medical school, they still went around acting fair-haired and chosen. This can be quite galling when you're not feeling very pretty yourself, in a shapeless gown with an IV in your arm.

The next day they peered at an x-ray and decided there might be a slight congenital defect in the heart valve, which sometimes produces aphasic blips of the kind we had experienced, and for which they prescribed one baby aspirin a day, I think to thin the blood. The tiny orange tablet looked almost laughably cute among the heap of medications. It was the first hospitalization that turned out to be neither ravaging nor time-heavy, and the second night we were feeling celebratory, telling all our friends on the tenth-floor staff about AZT. I remember walking with David Hardy to Immunology so he could spring me a few days' worth of drug, and as he unlocked the cabinet that held the elixir and all the protocols, I felt as special as if I'd been invited into the principal's office.

The day I went to pick up Roger was also the day of the office move from Century City to Kings Road. I was up early and directing the movers which files went in the back bedroom and which into the storeroom under the house. Roger's big walnut desk and credenza were put up on blocks in the garage, and a few days later I'd be frantically covering them with plastic trash bags, against a Manila rain that poured through the leaky garage roof. Except for three or four cartons that lay in corners in the living room and didn't get unpacked till six months later, I managed to fit all the files in the back bedroom so it could serve double duty as an office. Roger came home that day strong and optimistic, and it was just as well that he hadn't had to witness the move itself. With all his work in arm's reach again at last, the worst of the change was over.

That night I had a call from Joel, who told me Leo was no longer living with him in Santa Fe but was now in the spare room at his sister's house in Hollywood. Leo was still on suramin—four months after the doctors at UCLA had stopped the protocol there—and was also taking another drug IV for the CMV infection in his eyes. Finally I broke the news to Joel that Roger had AIDS. Then I talked to Leo and went into automatic overdrive, full of the bracing data on AZT. When I asked him how long he'd be on the CMV drug, he said, "For the rest of my life." It did not sound like a long time.

"Leo's better off there, really," Joel said before he rang off. "He doesn't really want me around. He wants me to get on with my life."

Cope prescribed twenty-five milligrams of Xanax for anxiety, and the oval white pill got added to the daily pile. Rog never experienced anything again quite like that sudden overload, and even at the very end, when he did have flashes of disorientation, he never lost a word. I was anxious myself as a matter of course, though it never struck me quite as deep because I didn't bottle it up so much as pour it out. I think back on the blank attack now as the one break in Roger's stoic persistence, a man who would not otherwise blink at fate. *Please don't let him lose his mind*, I remember thinking, even as the alarm died down and every passing day restored him to his wit and quick engagement. The fear was right, but the prayer was wrong. He had something else to lose.

In the first week of March we were both working again, stepping over each other in the two-tiered home office and fielding one another's calls. Rand Schrader dropped by one evening to visit, and as I was walking him down to his car, he told me a friend was staying at his place till he got his bearings. I knew Doug, his friend, had just been through a bout of PCP, but he was young and had managed to qualify for AZT with only one infection. Doug was still early as the pink of dawn compared to us. "But he's all right, isn't he?" I asked.

It wasn't a matter of his corporal health, said Rand. Doug had come home from work the other night to find that his lover of four years had moved out during the day, without a word, without a note; they never spoke again. Rand hadn't wanted to mention it in front of Roger. A year and a half later Doug is still going strong on AZT, and the man who fled, a lawyer, is still nuzzling about in the legal gutters of L.A. Certain people have cut him dead, of course, but it's a big city out there, and a man can manage to leave no traces of the lovers he's left behind.

We decided Roger was well enough to hazard a weekend trip to Palm Springs, something his parents had hoped for all winter, especially since the Israelis were still in residence in a condo across the pool. We set out Saturday morning, the first overnight outing away from Kings Road since the trip to the mountains half a year before. As it happened, the weather turned brutal, and when we came into the desert through a high-gust torrent of rain we somehow missed the turnoff we'd been taking for ten years. Then we got caught in a sandstorm that forced us to the side of the road, beating against the Jaguar like a steel band. We ended up in Banning, an hour out of the way, with me in a phone booth shouting to be heard above the wind. Roger had to vomit once in the ditch beside the road, while I tried not to yell. We arrived in the Springs under the weather, to put it mildly. Roger took to bed for most of the day we were there and scarcely ate. We all braved it out, being what family we could for each other. I felt schizoid as I veered from the front-room chat and the feeding rituals to check on Rog and wish we were home.

This will sound crazy, but we promised as we headed back to L.A. that we'd be down again the next weekend, when Jaimee and Michael and the kids would be flying direct to the Springs from Chicago. Somehow we chalked up the sandstorm weekend to Roger's being off AZT. Since the two incarcerations over the white blood count, Cope and the protocol people had grown more shrewd at stopping the drug when the white cells started to fall. This seemed to allow the count to recover again more quickly and, most important, didn't require protective isolation. Roger was off the drug during the ten-day period that straddled the two weekends in the desert, and we were either too naive or too preoccupied to worry that an infection might slip in during the off time.

In between, we had a busy week working at home, and Roger was a good deal stronger. We decided everything would be fine as long as we avoided sandstorms. Midweek I was down with a two-day spell of diarrhea, and again I sealed myself off in the front bedroom, wearing a mask whenever I left it. I didn't bother with stool samples and the attendant lab terror, though the current bout was as propulsive as the previous September's. This time I refused to knuckle under with hysteria and grimly waited it out. Yet I wonder if, as I focused on myself for those two days, Roger was experiencing any symptoms we didn't track down.

One night I went out late for groceries, since my schedule was firmly rooted now to a 3 A.M. bedtime. In L.A. you can do all manner of things in the middle of the night, if you don't mind the vampire pallor of your fellow insomniac shoppers. I came home and went in to Rog, who woke up and said, groggy and melancholy, "I just had a dream about Paris. Oh, I wish I could see Madeleine again." We hugged in the dark and talked about Paris, and what it would mean to plan for another trip, if only the AZT would give us a wide enough window. "We'll get there," I promised. "You'll see."

Since Christmas we had been making do without an attendant from APLA, and we were proud of our independence. One of the pleasures of normalcy was that Beatriz was coming on Tuesdays again. She'd been cleaning house for us as long as we'd lived on Kings Road, and we'd watched her progress from a shy girl with fifty words of English to a savvy and vivid woman, comfortably bilingual, with property of her own in Mexico. Because I worked at home and was thankful for the diversion, Beatriz and I would always gossip on Tuesday afternoon. She was very close to her brother Lorenzo, who'd arrived in the States from Guadalajara before the rest of the family and helped all the rest get oriented as they came across the border.

Lorenzo had been sick on and off for several months, including an extended stay in the hospital before Christmas, for diarrhea and general malaise. Beatriz and I had never used the "A" word about him, any more than we had about Roger in the year that had passed since the verdict. But though Lorenzo's diagnosis was longer in coming, due to the nonspecific nature of the pre-AIDS symptoms, Beatriz and I would talk whenever she came about the state of research and the elixir Roger was on. The medical terminology was difficult for her, of course, but she listened with absolute concentration, sounding out the Latinate names. We spurred each other on with optimism and spoke often about how changed we felt about the acquisition of objects. Whenever I see people's collections set out on étagères, with the price tags barely removed, I think of Beatriz dusting and shaking her head.

Friday before the second desert trip, Roger's friend Tony Smith from Boston flew down for an overnight from San Francisco, where he was on the last leg of a trip around the world. I could hear the cold in his head when he called from San Francisco and asked him not to come. But Roger wanted to see him so badly, and Tony swore he'd be assiduous about wearing a mask, so I relented. The two of them had a marvelous afternoon together, and dinner was served in a way that was second nature now that the white count had become a red flag. Though I would eat with Rog at the dining room table when we were alone, if we had guests I'd eat with them around the coffee table in the living room, while Roger ate in state in the dining room, ten feet away. I was especially vigilant about this arrangement because of Tony's cold, but I never stopped worrying that he'd lift his mask and blow his nose. And when the chaos fell full force the following week, part of me never stopped blaming Tony, as if the germs he'd carried from India or Micronesia were to blame.

The second weekend in the desert wasn't noticeably better than the first, especially for Roger. With Jaimee's family there the cast had doubled, and Roger made a real effort to be up and about—hunched over a bit and woozy, but brightening in the charged air of the children's wall-to-wall intensity. Those two nights we stayed at Rita and Aharon's place, and I stayed up late talking with them, because they were night owls like me. I also recall taking separate walks with Jaimee and Michael, where I stressed over and over how well Roger was compared to November and December. I felt as if I had to keep up everyone's spirits, and was convinced Roger had put out too much energy for Tony, and that's why he was so tired. Then, after I'd reassured them all, I'd go into the bedroom and sit on the twin bed and watch Roger sleep, trying to calm myself with his peacefulness as he lay curled in a spoon, a half-smile on his face.

After two ten-hour nights of sleeping in, Roger appeared to have proved me right, for he was much perkier Monday morning as we all sat at breakfast. Actually, as I remember now, Roger got up even later than I and was having breakfast himself, while the rest of us hovered and watched him eat. As he finished his cereal, he said almost offhandedly, "My eye feels funny." Immediately I was alert, but I casually asked him to elaborate, not wanting to alarm the family. "It's like there's a shadow in it," he said, blinking as he passed his hand back and forth in front of the right eye.

Though he shrugged it off, I said we'd call Kreiger when we got back to L.A., and the worry dissipated in the round-robin of family cheer as we made ready to leave. Michael, a rabid Cubs fan, gave Roger an umpire's cap from the National League for luck. Then all the way back to the city, I kept thinking of Leo on intravenous eye medication "for the rest of my life." A new drug that had come on line in recent months to battle cytomegalovirus was one of the few bright spots in treatment. Previously CMV had rendered a lot of

AIDS patients blind in the early years of the calamity. Then I started obsessing about the cotton-wool patches that had floated benignly in Roger's retinal sky for a whole year now. Had one of those clouds begun to darken?

Beside me, Roger kept squinting, and I asked if it hurt or was getting worse. No, he said, but the squinting didn't stop. When we pulled into an off-ramp Denny's for a bite of lunch, I called Kreiger's office at UCLA in a panic, but he was away for the day and his service was picking up. As soon as we got back to the city we had to retrieve the dog at the kennel, and he was hysterical. So we had our hands full unpacking and settling in again, and Roger needed to rest from the trip, especially since his fever went up that night. We put off the eye till the next day.

I don't recall if Roger's vision was worse on Tuesday, but the fever was persistent, and now I was certain he'd picked up some kind of flu from Tony. Cope said that was entirely possible and told us to monitor the fever and check in by phone the next day. He knew how reluctant we were to come in for no reason at all, especially after the recent false alarm. As for the eye, since Kreiger would be out of town till Thursday, we made an SOS call to our ophthalmologist friend, Dell Steadman. He met us during his lunch hour at his office in Beverly Hills. As he gazed into Roger's eye with his scope, I held my breath the way I used to do as a child whenever we drove by a cemetery.

No, said Dell, there was no CMV in evidence. The cotton-wool patches were stable. And since he could see no other problems, he suggested the optic nerve might be temporarily damaged by a flu or cold virus. Roger's vision in that eye had dimmed some more, but not dramatically. We went home relieved and tried to forget it, tried to go back to waiting for AZT, but the fever wouldn't go away, so the next day Cope suggested we'd better come in.

The rest of the week is a blur of apprehension and horror. Kreiger and two other eye doctors examined Roger over and over, and though the business about the optic nerve made sense at first—by now Roger was *seeing* mostly shadows out of that eye—the retina began to show subtle signs of damage. Suddenly Kreiger wasn't satisfied. I could see he was puzzled and thoughtful, even as he concurred for a while that the vision would surely return—or perhaps he just neglected to contradict our own tense optimism. I don't know when he decided to put Roger on a high dose of acyclovir, the herpes drug. By then I was on the phone nonstop, trying to field all the info I could find about AIDS and the eyes. The problem was, there was no way to be sure it was a herpes infection, because you can't do a biopsy of the eye, except by autopsy. Yet Kreiger decided to treat it as if it were herpes, though he'd only know that he guessed right if the forward creep of infection stopped.

I don't know myself what I was trying to find out with all my phoning—any anecdote would do, it seemed, as I pieced together a nightmare collage. I remember talking to a man who didn't know who I was, whose number was given to me by one of the Tijuana mules. He had gone blind only a few weeks before and was still choked with sorrow about it, yet he bravely told me his whole story—the misdiagnosis, the prolonging of treatment till it was too late, the breaking of the news, the blackness.

I told him I was sorry and then about my friend. Yes, he said, he understood; his own friend had died just after Christmas. Among us warriors there is a duty to compose ourselves and pass on anything that might help, no matter how deep the grief. Two weeks after Roger died, a frantic acquaintance called to ask about the meningitis drug that hadn't worked for Roger, who died with it in his veins. Just the mention of the word took my breath away, as I answered questions about the convulsive side effects. But I thought of the blind man trying to help me save Roger's eyes, and so I stayed on the meningitis case till the crisis was past.

By week's end the vision was effectively gone in the right eye, though Roger could still distinguish light from dark and make out the shape of my hand as I passed it back and forth like a metronome.

Roger's parents had come up from Palm Springs, and they sat with him while I made calls from the corridor phone. We managed not to panic because the fevers had passed and Roger was feeling fine, and no one had yet told us the loss was irreversible. All my scattered research among the blind kept coming back to CMV, which wasn't our problem. No one seemed to know very much about herpes in the eye, but at least it was treatable, everyone said.

So I don't know which came first—hearing the infection had spread to the good eye, or hearing there would be no return of what had gone. It took a day or two for the acyclovir to kick in. The infection finally stopped moving, and at last Kreiger was satisfied he was dealing with herpes. Now he suspected that several other cases of blindness he had heard about lately were herpes-related. Yet they were being diagnosed CMV, and the vision evaporated for want of the right prescription. The blackness was, in fact, preventable. In a matter of days the infection had managed to destroy the retina of Roger's right eye, burned like chaparral in a canyon fire. I thought of the blind eye in just that way—as scorched earth, the retina itself spent and insubstantial as a flake of ash.

We must have all been in shock, but if there was ever a time we had to see the cup as half full, it was now. We poured all our emotions into relief that Kreiger had been so smart and so tenacious, because if he had let it go so much as another day or two, Roger would have been totally blind. Jaimee remembers Roger saying a week later, when she and Michael were visiting on Easter Sunday, "I'm willing to give them one of my eyes, as long as I can keep the other one." So by then Roger had bought into looking on the bright side, a process his parents and I were eagerly, almost frenetically, engaged in. His vision had shrunk by that point to a cone without much peripheral reach, but Kreiger and Cope assured him the eye would adjust and compensate, and there was no reason he wouldn't be able to drive a car again.

But before that affirmation took hold in him there was an implosion of despair, where he nearly slipped away. He seemed to be processing it all reasonably well, communicating with us, connecting still to work, reassuring *us*. There was no way to gauge how much he was simply trying to make us feel better, for that was his instinct always, conscious or not. All I know is that Dr. Martin the psychiatrist went up to the hospital for a session with him, and when he got back to his office he called me and said, "Roger was very confused today. You do realize there's brain involvement."

I don't recall there being an instant's time between hearing that phrase and driving the car at fifty-five down Sunset to the hospital. When I got there Roger was asleep, but I coaxed him awake. I wasn't sure how lucid he'd be, and scared of the distance, but I was determined to find a way in and find him a way out. He was all there, I could tell right away, but very sleepy, almost drugged. I felt this urgency to keep him talking, the way you would try to keep a man conscious who'd taken an overdose of sleeping pills. Yet for once my urgency was unaccompanied by hysteria. Over the next several hours I was absolutely clearheaded as I kept him engaged and lobbed him questions, never overtaxing him and always keeping calm because I sensed he was being dragged under by his own anxiety. I wanted to make the real world, here with me, the easy one. I understood intuitively that the great tiredness was a kind of shutting down. I knew we must stay in absolute sync, for the enemy had grown so subtle, its camouflage so chameleon, we had to be on constant watch.

That night when Cope came in, I remember, Roger was quite without affect, picking at his food and peering suspiciously with his one eye. Cope couldn't seem to get anything from him but monosyllables. Roger was still having pain from the shingles, and he'd clutch his side when he sat up, as if stabbed by an old wound. We'd both heard enough good cheer about how thrilling it was to be left with two thirds of an eye. But Cope stayed and persisted, talking to me instead, at one point asking me how I was sleeping. Okay, I said, having chipped my way down to half a Halcion and a quarter of a

Xanax. I was compulsive about keeping my milligrams low. And suddenly Roger cracked up, laughing at the thought of me quartering the little white pills, he who had borne witness to a thousand of my small compulsions of hygiene and general nest behavior. Cope seized on Roger's laugh, and in a moment we were all chatting again, the three of us against the calamity.

I think of that as the moment when Roger came back, though there were a few days on either side when he was closed up like a flower. "Came back," I say, in echo of the highest honor anyone's ever paid me. It was a few months later, and Roger and I were having a walk late at night, and I was fretting about being not strong enough, when he said in disbelief, "But Paul, you were the one who brought me back."

That was the day he meant, when I talked him out of "brain involvement." But even back in November, during those grave days when he almost died, as we waited for the AZT to come in, Roger had gripped Gottlieb's hand one day and said, "Bring me back. I'm not ready to go yet." So that was always the way we thought of it, going to the brink and coming back. When I recollect the times we made it through the dark, I remember feeling as if I were pulling him in from drowning, out of a whirlpool, then breathing the life back into him. As long as he knew who we were, he was here.

We were much further off than the moon in the weeks we fought for those millimeters of sight in the left eye. It was all a hair-trigger waiting game, as they tried to figure a dose of acyclovir that would let him come off IV but would hold the virus. It seemed like agonizingly delicate guesswork, but the longer it went on, the more ground we regained from the enemy. Meanwhile various bulletins intruded from other fronts in the war, but they were all the old bleak stuff of pneumonia and lymphoma, lesions and dementia. We were in a territory not yet on the map, and the herpes battle seemed yet a new intractable dilemma, another in an unending series of dead ends.

Once more I was on the swing-shift schedule of being in the hospital afternoon and night, with a break from six to eight to feed Puck, check messages and drop by the gym for twenty minutes on the Lifecycle. I played the messages now with a sort of flinching of the will, as if I couldn't bear it that somebody might want something. I remember a message from Joel: "Leo's not going to make it this time, and I'm staying until he goes." He left a number, which never answered, and by the time someone picked up, three days later, Leo's ashes were off to West Virginia, and Joel had gone back to Santa Fe. I kept it from Rog for several weeks, because I didn't want him connecting up Leo's eye problems with his sudden demise.

But I must have made the connection myself, and was processing it unconsciously. I was talking to Craig recently, reporting about a friend who went blind in one eye, with no warning. Craig volleyed back the tale of a mutual friend in New York who'd lost all but peripheral vision in one eye because his doctor rescheduled an appointment, putting David off four days—four days that stole his outer edges. And we realized we have this doomed sense now that when the eyes begin to go the brain isn't far behind, that eye problems are the break in the central nervous system's defenses.

Four friends got together and bought Roger a pyramid-shaped talking clock, an eccentric but useful gift that proved a delight to show off, due to the Japanese accent of the voice of Time. Roger always loved gadgets that masqueraded as toys. On several occasions he followed the demonstration of the clock with a story of squiring Borges around Cambridge, when the old fabulist was visiting Harvard and Roger was in Comp Lit. Borges had a big pocket watch that he read with his fingers, but very discreetly, as if he didn't want to seem rude, like someone who watches the clock. Then Roger would quote the Borges line in which he speaks of the irony of his blindness: "I who always thought of Paradise as a kind of library."

One Saturday afternoon I had to be back at Kings Road at three, because I promised Craig I would deliver his next mule shipment of ribavirin to Paul Popham, who was out from New York for a long weekend. Paul had been diagnosed in the same month as Craig, Bruce and Roger, and now thirteen months later he had a bout of PCP behind him, plus he was buckshot with KS. He arrived with his lover, Richard, a wartime bond for sure. Paul having lost a lover to AIDS in '81, they always knew, as Richard remarked at Paul's memorial service, that the nightmare might consume them at any moment. Paul looked thin but tough that afternoon, and I could faintly detect the skilled makeup on his neck to hide the lesions. Richard was undiagnosed, yet he looked terrified and said almost nothing. I was too shy to engage him about the similar roads we were traveling. A month later Richard was diagnosed with lymphoma.

Paul expressed his angry condolences about Roger's struggle with the herpes, then declared, "I'll fight this as long as I can." But he said it with a shrug that wasn't afraid to be hopeless and overwhelmed. This from a Vietnam War hero who testified at Congressional AIDS hearings in Washington when he could barely climb a flight of stairs, pleading with the government to notice us. I love that fatalist's courage—a courage that has cold reality and a sense of the tragic built in.

The hospitalization for Roger's right eye was three weeks long, and after the first wave of furies—shocked, numb, resigned—we made our bargain for what was left and put the crisis behind us. Herculean denial, perhaps, but there was a genuine air of relief as we went back to the space capsule mode, adapting our real life to the confines of 1024. Besides, Roger was back on AZT, and that never failed to point us forward, a little more battered but single-minded as ever. Meanwhile Bernice was able to channel a lot of her own anxiety by volunteering to be Roger's secretary, and the two of them began an ongoing process of filing and making calls, taking care of a babel of details that had gone unattended in the previous weeks. With Roger busy working in the afternoons, I stooped to pick up the thread of my work with Alfred.

Joe and Stuart from Philadelphia had recently sat down with Roger to do their wills, and now he dictated drafts of both over the phone to a word processor in Century City, who agreed to moonlight for him by the hour. Another client went to the hospital to have some business papers drawn up. Roger was working productively and comfortably now with a co-attorney, Esther Richmond, an old friend and fellow sole practitioner. With extraordinary adaptive skill he had drawn together a system that covered his bases and still met his own exacting standards. In the process he preserved a small corner of the market that was his, and he kept it alive till the end.

Yet I recall how hurt Roger was when he lost his first big client for no other reason than AIDS. He was a surgeon from Orange County, a fresh-minted millionaire surrounded by business managers, pulling in the mid-six-figures. He'd been a client of Roger's since the law firm days. I always used to think of him as Roger's one normal client, who just had money and not a lot of problems but needed a lawyer to handle his byzantine corporateness. Roger never missed a deadline with him, never neglected to return a call throughout his illness. But Dr. Orange was aware that Roger had been hospitalized several times and finally asked what was the matter, was there anything he could do. Roger shrugged and bit the bullet: AIDS, he said, but hastened to add a positive word about AZT. The doctor made all the right clucking noises of sympathy, and three days later a lawyer called and said he wanted all of Orange's files forwarded to him.

When Roger queried the doctor, Orange swore he'd been misunderstood and quickly backpedaled. "Of course I want you to finish what you're doing for me now," he said. "I only wanted to set up a smooth transition." Beware of transition, the euphemism that kills without leaving a mark. Jaimee ruefully told her brother that Orange's new lawyer had obviously hustled to fill the vacuum, but that Roger's work wasn't dependent on one client, so let it go. Yet Roger had a hard time over the whole issue, seeing it not really as AIDS discrimination but as if somehow he'd failed as a lawyer.

By now Rita and Aharon were on their way back to Jerusalem,

with a gravely ironic and casual good-bye as they left, and a few days later Al and Bernice headed home to Chicago, saying they would be back the instant we needed them. As we gathered ourselves together to leave the hospital we were eager to get a new prescription for Roger's glasses, so he could have full use of the good eye. Our friend John Orders, on a year's sabbatical from CAL/ARTS, came by one morning and took the prescription downtown so it could be processed the same day. About 8 P.M. Roger had his new glasses, and smiled with pleasure at how well he could see. In that moment—the satisfied smile as he gazed around the room, taking in the world again—he was whole once more. It may have been for the last time, but he glowed with possibility, ready to leap back into life without a trace of bitterness.

There was a last bedside exam by Kreiger, who said he would monitor the eye every week or ten days to make sure there was no recurrence of infection. Meanwhile Roger would be on a high oral dose of acyclovir. Almost by way of an afterthought, Kreiger mentioned that the only other problem he could foresee was a detached retina, since there had been sufficient damage to the good eye to weaken the connection. But he assured us the probability was remote and we shouldn't worry about it.

Roger came home the next day. We had a temporary nurse's aide from APLA to help with the transition, but were hoping we could make do on our own within a few days. The nurse was a middle-aged black woman who'd buried her husband the previous year from cancer: she quietly read her Bible when she wasn't helping Roger, highlighting the text with a yellow Magic Marker.

I don't even remember any wariness in us about the homecoming, perhaps because Roger had been stabilized in the hospital for several days on oral acyclovir. In addition, the AZT was having its noticeably revitalizing effect on Roger's strength and alertness. John Orders dropped by Friday morning with some groceries—I was still asleep—and he sat and visited with Rog in the pool bedroom. We'd known

John since Boston, met him on a sunny spring day on the Esplanade, walking with a friend who's dying of the plague now. On this equally bright spring morning ten years later, John and Roger were happily chatting and making puns when suddenly Roger tilted his head and said, "It's awfully dark in here. Do you think it's dark?"

"No," replied John in an ashen voice, feeling, as he told me later, a terrible sense of dread.

I woke up shortly thereafter, and Roger told me—without a lot of panic, almost puzzled—that his vision seemed to be losing light and detail. I called Dell Steadman and made an emergency appointment, and I remember driving down the freeway, grilling Rog about what he could see. It seemed to be less and less by the minute. He could barely see the cars going by in the adjacent lanes. Twenty minutes later we were in Dell's office, and with all the urgent haste to get there we didn't really stop to reconnoiter till we were sitting in the examining room. I asked the same question—what could he see?—and now Roger was getting more upset the more his vision darkened. I picked up the phone to call Jaimee, and by the time she answered the phone in Chicago he was blind. Total blackness, in just two hours.

He didn't cry out, not then. He was too staggered to howl like Lear, and all I remember is a whimpered "Oh," repeated over and over. Then Dell came in and examined the eye and said as calmly as he could that indeed the retina had detached. As the two of us choked on nothingness, he put in a swift call to Kreiger, and they talked about scheduling an immediate reattachment. Dell had nineteen other patients waiting, and there was nothing else he could do. He said he was sorry and left, looking helpless. We sat there stunned, clinging to each other's hands. I think I tried to pull out of it and focus on the operation, but neither of us could think at all as we tottered forth from the suite, me leading my friend as he groped a hand in front of him. The nurses' faces were tight with pain.

I don't know what we said to each other. I think we just numbly went forward—I had to hold him close and lead him down into the parking garage, then somehow get us home safe through murderous Friday traffic. I made consoling noises, but they made no sense. When we got back to the house I settled him in the bedroom that two hours before he could still see. The nurse tried to make him comfortable, but still that frail and broken "Oh" was all he could say. I called people for him—his parents, mine, I don't remember who—and at last he let the cry tear loose. "I'm blind," he wailed as he clutched the phone, again and again, to everyone we called.

None of the meaningless, unsolicited consolation that people have murmured since then—about the logic of things and desirelessness and higher powers—will ever mute a decibel of that wail of loss. I had to force myself to stand my ground in the house and hear it, and not go mad or dissolve in a tantrum. Everybody he talked to cried with him, but I was too scared to cry. Besides, I had to get us through to Tuesday, for Cope had called right away to tell us the operation was scheduled for Tuesday morning. I listened in on his call to Roger, huddling in the shade of his compassion, trying to learn what it was people said when the worst had happened. He listened to Roger's woe and terror, *really* listened, with an "Oh" that echoed Roger's own. Then he spoke and gave comfort and made us hope. All through the calamity I've heard the noblest people do that: Somehow they find the words.

It was Joe Perloff that Roger turned to the next day. I greeted Joe and Marjorie at the door, and behind me Roger felt his way along the hall and came wide-eyed into the study, as starkly blind as Oedipus, struck down and gaping with the horror of it. He broke down crying as he clasped at Joe, and they sat to talk while I went dazed into the living room to sit with Marjorie. Joe spoke to Roger about the fear of heart patients before surgery, and he said the only wisdom he'd learned from them was how they took the enormity of it one small step at a time. Since Joe was also Kreiger and Cope's colleague, he was the perfect bridge of security that day, anchoring our trust that we had the best on our side.

Not that either of us was capable of feeling much better. I kept the calls coming in from friends and family, and a stream of visitors Saturday and Sunday who couldn't think what to say but who came. Anything to keep Roger from sinking into himself, now that the world had cracked in two. I remember Rog on the phone with my brother on Sunday night, the natural empathy between them because of my brother's handicap: at least Bob was someone to turn to who knew how little anyone understood.

I don't think we even ventured out of the house all weekend, except to sit in the yard. Dr. Martin was extremely helpful, focusing us on the sensory deprivation, giving Roger clues to bring the world back—music on, think of the plate as a clock, organize the bedside table, sit outside, keep talking. Monday Roger's parents arrived from Chicago, scarcely having unpacked from the trip home. I took Rog over to UCLA and checked him into Jules Stein Eye Institute, a whole separate building at the medical center, sheathed in rosy travertine, lavishly appointed throughout.

Roger and I had grown accustomed to the tenth floor and its staff, so much so that we thought of it as an annex of our lives. The third-floor inpatient facility at Jules Stein was entirely different. They had only the one floor of rooms, and most were empty. Either the patients they dealt with were in and out, or they were pricing themselves out of the market. It was clear to me from the nurse's intake interview that they'd never handled an AIDS patient before. When she started to say the wrong thing about isolation and then wouldn't agree to wear a mask to protect Rog, I began to get sharp, till Roger begged me to stop. I came within a hairsbreadth of being kicked out on my ass.

Kreiger came in and explained the operation, and we found out the FDA was screwing up all over the board, not just in the AIDS department. They'd refused to approve a form of silicone used successfully for a generation in Europe to reattach the retina. In order to obtain some for Roger, they had to spring it from an animal study in northern California, since the silicone was allowed in veterinary medicine. We didn't know till afterwards that the operation had never before been successfully performed on an AIDS patient. Kreiger was cautiously optimistic about the chances of regaining full sight. Roger had been his patient long enough that Roger trusted him, and Kreiger's air of self-possession was incalculably calming.

Later Cope made his rounds to wish Roger luck, and broke the news that he'd be out of town at a conference for the next few days. "When I come back, Roger," he said, "you'll see me."

Jules Stein was a pretty forlorn place the night before the operation, with only three or four patients. Roger's parents and I stayed close around him, conversing and somehow keeping one another from spinning out. I stayed very late, went home and didn't sleep, and was back early in the morning, just after they'd shaved his eyelashes. I'd brought up a pile of mail, and as we waited for them to take him down to surgery we paid some pointless bills. Two orderlies arrived at ten-thirty and bundled him onto a gurney; his parents and I walked beside it as we headed for the elevator. Strange, how in all my memories of these days I keep forgetting Roger couldn't see us, guarding him as we descended to the operating room. I think it must have been Al we had to thank for the conversation that kept us going, all the way to the swinging doors. Then it was time for us to leave him, and Roger looked up at us—couldn't see, just looked up helplessly and started to cry. I froze and couldn't think what to do, but his mother bent close and said, "Roger, take a deep breath." We squeezed his hand and they wheeled him away.

Nothing has ever been longer than the next three hours, not even the day he died. Al had made a plan that we would meet Sheldon at the Bank of Los Angeles and go out for lunch to the Deli Diner, a fast-food place Sheldon owned on Third Street. I would have been content to just sit upstairs going crazy, but obviously the parents had the idea we ought to keep distracted. At the bank, Sheldon was forcibly cheerful and introduced us to his new assistant, Len, coyly remarking to me behind his hand as we went into the Deli Diner that

Len was "even more gorgeous than the last one, don't you think?" I did not hazard an opinion. By now I was used to him saying the wrongest, dumbest things imaginable, anything to avoid emotion.

I was feeling more and more unreal, it being the first time I ever waited out surgery in my life. You almost go into suspended animation, scarcely breathing, like the aura before a seizure, everything physical heightened. This is all fear, but the craziness of it is worse in public places. Sheldon's restaurant partner, Don, personally served us and tried to say something cheery about Roger: it was the one time Al and Bernice turned away and swallowed tears. Everyone forced me to eat. Sheldon didn't go back with us to the hospital, nor would he visit Roger throughout his stay at Jules Stein. Something about the blindness freaked him out, touched a nerve he couldn't control. My rage and contempt toward him were boundless all that week.

Kreiger had said he would call us upstairs as soon as he came out of surgery, and by the time we returned, Roger had been in for two hours. There was no one else in the waiting area or walking in the corridor. It was as if this wing existed just for us, which only made us seem more separate and more lost. I told Al and Bernice I had to go for a walk, then went out and tramped the UCLA campus for a half hour. The main thought in my mind throbbed like a migraine: If the operation didn't work, then I would have to be the one to break it to Roger that he was blind forever.

It was a flawless California day, bright and cool, the sky Della Robbia blue. I stood on the hilltop terrace outside the library, with the green plain of the playing fields below and the Bel-Air hills to my right on the western ridge. *He'll lose all this*, I thought with a pang, *and then I'll be all the eyes we have*. Always for the two of us the window on the sublime was the eye in nature or the eye in art, a seeing refined by twelve years of wilderness and museums, till we saw certain things exactly the same.

I hurried back to the medical center, trying to screw up my courage, but still his parents waited dull-eyed in the third-floor lounge. We sat silent for another half hour, and I was around the corner by the pay phones when the call came through at the nurse's desk. I froze inside as I heard her beckon Al. Then I walked around to join them, whatever the verdict was, and heard Al say: "Yes, Doctor, yes. Oh, thank God." Bernice and he were huddled together over the phone, crying, and I put my arms around both of them, the tears breaking from me at last, all for joy. There was nothing before or after quite like that flood of relief—calling the family, spilling the good news, then waiting impatiently for Roger to come up from Recovery. About an hour later the gurney appeared, and I saw the great bandage over his eye like a battle dressing. As they passed me I leaned down and said, "Rog, it worked, the operation worked!" And he smiled, still a bit groggy: "Oh, good."

"Oh, good." If every minute of the nineteen months was worth the struggle, reclaiming a corner from death, some moments were the most exalted of my life. The sheer gratitude after coming through fire is so profound, so first-things-first, it makes you laugh inside when people say you are brave. And there wasn't a shred of anxiety in Roger as we started up the mountain again. He stayed in the hospital for about a week, encouraged daily by Kreiger's satisfaction as he eyed his work. Then came the sensation of light again when the nurses changed the dressing. Bernice and I had to watch this procedure closely, as we would be changing the bandages for ten days after he came home, cleaning the eye and putting the drops in.

The protocol for the dressing change was an enormous responsibility, but I champed at the bit to learn it. Finally, something to *do*. And when you do this part you come to see there's something nearly sacred—a word I can't get the God out of, I know—about being a wound dresser. To be that intimate with flesh and blood, so close to the body's ache to heal, you learn how little to take for granted, defying death in the bargain. You are an instrument, and your engine is concentration. There's not a lot of room for ego when you're swabbing the open wound of the eye.

Indeed, we were all so busy afterwards, and calming down from the heart-pounding fear of the four days preceding the surgery, it began to seem the merest technicality that Roger was still in the dark. Basically, once the operation was safely past I didn't see him as blind anymore. On the contrary, we were engaged in the process of restoration. For once we were going to get something back. And if that seems like more denial, wildly out of phase with Roger's continued imprisonment, even he didn't start to despair till a few weeks later. By which time the heightened urgency had given way to a slow recovery of the blur of the world, detail by agonizing detail.

But I brood still about how it must have felt, blind all those days, even with the hope turned up full volume. He would turn his head now when I came into the room talking, reaching out with his ear instead of his eyes, which had always been searching and playful at once. Blue and mild and deep as Walden Pond, which Thoreau called "earth's eye." The subtlety of a thousand looks, the range of expression so different from the starkness of the turned head, the cocked ear. And then what it must have been like to be walking outside on somebody's arm, or indoors feeling his way from room to room.

I don't mean to insult him with pity. He was brave and resourceful, stubborn and gallant, and his blindness left him in a temple as clear-eyed as Borges's library. But it was also his buoyant stride and ease of motion that were compromised, he who had such a hunger to be out in the world. He'd always preferred to walk, never cared how far away we parked, we'd get there soon enough, and meanwhile there was the pageant of the crowd to see. He'd spent three years in Paris doing what bohemians do, walking around and sitting over coffee, all the better to people-watch. The street was where he contemplated life, part of why he was so plainspoken and had no airs, preferring a clean, well-lighted place to three stars any day. I can't imagine what it was like, that growing sense of the world stolen.

Perhaps you compensate here as well. Through all the long convalescence of the eye, Roger had an immense patience, trying to outwit the blur and stealing things back by fragments. And if his liquidity of movement had been snatched from him, he was quick to concentrate the remainder of his power: talking and listening. In that regard what gave him the most coherence was the work he was doing with his mother. He'd sit with her in the living room for hours as she read aloud documents and the mail, with Roger directing her where to file and whom to get on the phone. As I puttered around the study, it was such a pleasure to hear the rhythm of things being put in their places.

Since he was feeling strong from the AZT, the main medical business of the day was the change of dressing, morning and night. And always the ritual question: *What do you see?* Within days of returning home he could make out the shape of us standing over him. Then a week later, when the shapes acquired edge and detail, color began to seep in again. I remember him seeing the red in the Erté print on the opposite wall.

His parents spent the whole day with us in the house, but were living at that point in a furnished apartment in a building Sheldon owned at Doheny and Melrose. Sheldon had taken a house in Malibu with a boyfriend for May and June, and he thought the apartment would be easier for Al and Bernice, since there were workmen at the Bel-Air house, acid-washing the stone floors. This apartment business occasioned a good deal of fuming and friction in the family, because Al and Bernice had trouble with the laundry, and there weren't enough forks or a can opener. Sheldon wasn't good about any of it, annoyed to be bothered with details and insisting on the inviolability of his time at the beach.

Meanwhile Alfred and I had a windfall. A project was literally dropped into our laps, by a couple of producers we'd worked with in the past. It was the first thing we hadn't had to go out and scrounge for, and would provide a structured amount of work to do, a few hours a day for the next several months. So seductive are the old ways that Roger and I were soon ventriloquizing life before the calamity, when the two of us were taken up with work but always glad for the end of the day, when the pals regrouped.

Yet in the weeks that followed the operation, my most concentrated energy went into the garden. We hadn't planted, had hardly even watered during the whole last year. The trees were dense and shaggy, the shade so dark that only the dinosaur ferns had survived. The sunny patch above the pool was just a sprawling bougainvillea banked above a patch of arid dirt. When Roger was resting I'd go outside and weed and plant and water. In California you don't have to be very evolved as a gardener to fill the yard with green, and I'd report my progress as Roger slowly regained the light in May.

Or he'd sit over lunch in the dining room, which opens on a courtyard under the trees, and we'd talk as I knelt and planted impatiens. Then he'd come out with his father to see what he could see. I wanted to make a place for him to discover, one color after another, something concrete to come back to. On the cracked terrace under the Chinese elm I potted a bushy gardenia full of buds. I suppose it was hard for either of us to believe anymore that we were going back to the world out there. We hadn't been to a restaurant in eight months. So when I finished the gardening I had a new pool heater installed, then brought in an electrician to put temple lights under the shrubs. Here was where we would spend the good hours of the summer.

Our friends were wonderful to us. Ten or a dozen different ones stayed in regular touch, but also holding back a bit so as not to crowd or intrude, especially when Roger's parents were in town. After Al and Bernice left in mid-May, a couple of nights a week friends would bring dinner in, so there were things to plan and look forward to. It made us feel glad of the house and echoed all the happy times when we'd had people over in the past, when we had so much to celebrate.

One evening, I remember, Joe and Stuart were over, with a raft of high-toned takeout from Irvine Ranch. We all sat talking about the cruel and soulless posture of the Catholic Church on the subject of being gay, the Judas priests who hid their brotherhood with us. Stuart and Roger were discussing John Boswell's *Christianity, Social Tolerance, and Homosexuality*, a book they both loved, which details the benign attitudes to gay love in the early church, even to gay marriages. Roger had known Boswell at Harvard and recalled talks with him on the steps of Widener, filled with amazement at the young historian's erudition. As they talked of the late Roman world and the early Christian, I thought as I often did now—as usual, making history up as I went along—about the physicality of the pagan gods and the ancient men of our kind.

So we had a few weeks free of further calamity, and during that time enough of Roger's vision came back so that he could see his way around. We began driving up to Laurel Canyon Park again to run the dog. I remember the day in Kreiger's office when Rog identified the E at the top of the chart. The next month or so was a constant movement forward as he worked his way down that chart, a new line every week. Or he'd sit in front of the television and peer intently out of his good eye, picking up images. One night he listened to the Mozart *Requiem* on PBS, and the camera stayed in one place long enough so that he saw it as well as heard it.

We had a new attendant from APLA. Dennis was a black man in his early twenties, possessed of great gentleness. He could hardly boil water in the cooking department, but tried so gamely Roger grinned and bore the odd juxtapositions on the plate. Roger also made arrangements with a young man who worked a part-time sales job and wanted to moonlight helping people with AIDS keep their businesses going. With all our systems in place, I took an afternoon off and went to see Dr. Scolaro, jettisoning at last the Ferrari doctor. Scolaro was an activist who believed the virus was conquerable at an early stage, and he was thus aggressively studying all the combinations of antivirals. I found out my T-4 helper cells had shrunk from 590 the previous summer to 430. No drug of any sort had yet been found to increase the T cells. Once the virus consumed them they were gone. Scolaro prescribed acyclovir, the herpes drug, on the theory that it lowered viral activity in the body and perhaps muted some of the cofactors of the AIDS breakthrough. He said I should be having the numbers tracked every three months now, and if the T-4 number showed a bad trend downward I should probably go on ribavirin.

I know it was only three weeks of repose before the next crisis, but in memory the days of late spring are longer, perhaps because time was so precious, especially with sight returning. We were very laid-back ourselves, as I would read Roger the papers in the late afternoons or, at night, from *National Geographic.* One long piece I read aloud recreated Ulysses' voyage home from Troy to Ithaca. I gave elaborate descriptions of all the maps, to locate the whole itinerary on the ground of Roger's imagining. He couldn't read yet; reading still seemed far away. But it didn't feel in the least as if we were waiting for death or the next disaster. And we were so close now I couldn't think at all anymore without thinking first about Rog. In sum, we were doing the best we could with what we had left, and more and more it was like Diogenes tossing away the tin cup because he could drink with his hands. It turns out there is no end to learning what you can do without.

# ·XI·

People around us were full of a certain fatalism now. Not that they'd put it in so many words (if they did, I'd fire an immediate warning shot). Cousin Merle called from Oakland to say she'd like to come down and see us soon, "because we don't know how long it's going to be." I told her sternly we didn't talk that way, and no one was planning to die around here. Or one afternoon I was walking with Rand in the hospital, and he said how painful it was to have a friend "in the late stages of AIDS." That wasn't how we thought of it at all, I retorted coolly. "There *is* no early or late," I said. "There's only how much you fight it."

As I write this it sounds, even to me, as if I was living a total illusion. And I wonder if Roger felt it as strongly as I, that to talk about death at all was to leave a door unlatched. Jaimee and I were so bullheaded certain we'd beat it. I think she and I set the tone from here on, the held breath as we passed the graveyard. Though Roger would sometimes get snappish at us—"I don't need any more pep talks"—I never had the sense that he was any more eager than Jaimee or I to talk about the end. Perhaps he held it all in for us, deeper than he wanted to. A chill of guilt still shivers through both of us sometimes that we didn't let him speak. "What were you supposed to say about death?" Sam asks me now. "That it sucks? Don't worry, you all knew that."

The unspoken fear and sorrow in our friends were every bit as troubling as any remark that overstepped. Around May 15, Susan and Robbert left for their two months in Europe. They promised to send a blizzard of cards, and did, but we bid them a hollow *bon voyage*, as if we knew what a feat it would be to survive till they returned. Conversely, everyone close to us could see how extraordinarily Roger had survived the latest indignity, and every time another crisis had been navigated the hope would ripple outward. At least for the first month after the operation we left no stone unturned, as we fought to maximize what sight was left.

So there was probably something of a mood swing among our loved ones—resigned one minute, inspired the next. Sometimes the end must have looked to be inevitable and soon, and other times, when we defied the odds, our team was as exhilarated as we were. Al and Bernice had asked Cope in April how long it would be, and he shrugged and said, "Three months, six months . . ." None of them ever told us that, and when I finally heard it I could only think defiantly that we'd made it the full six. Yet it was just as well nobody tried out the odds on us. For a while there, we were too busy winning to lose.

Kreiger put us onto the Center for the Partially Sighted in Santa Monica, and we went for the first appointment early one morning at the end of May, shy and a little uncomfortable at having joined the disabled. Happily, the Center turned out to be a haven, and from the very first interview, with a painter named Joey, who was openly on the bus, we realized we were in the presence of even more pluck than we could muster on a good day. The operating principle was, if you had the smallest crack or shadow of vision they'd find you ways to see with it.

Joey turned us over to Dr. McAllister, an ophthalmologist, who was able to fit Roger with eyeglasses that took the blur from his field of vision. Now he could see things for real again, not well enough to walk in a crowd, not strong enough to read, but he saw my face and the temple lights and the first gardenias. A week later McAllister showed us a closed-circuit television that could magnify a text so the letters were an inch high on the screen, white on black. He lent Roger a lighted magnifying glass, better than the one we'd bought at Koontz Hardware, so Roger could scan a menu or the headlines. Also a pair of wraparound shades to cut the glare. "Do I look like *Miami Vice*?" Roger asked me, deadpan. All of these advanced his vision, and equally our morale.

The only difficult time I remember at the Center was the day we had to learn to walk blind. A soft-spoken Hispanic woman was our teacher, and she showed us the best techniques for a sighted person to lead around a blind person. Roger would grip my elbow, and we learned how to turn and negotiate narrow spaces and sit down. It was all presented in terms of the most rigorous practicality, completely unsentimental, grounded in retrieving some measure of the range of motion that sightlessness had stolen. Still, we were being taught to go into the world with a handicap, which is always a brutal transition.

Intolerance, of course, is common law in America. You start to realize that generations of the physically challenged were kept at home and turned into invalids, thus compounding their loss of the world, because they were thought to be aesthetically problematic. For this pitched battle I know I was helped by the early experience of my brother's disability. However difficult the memory trace of wheeling Bob in public, however white hot my rage when people would stare at Roger, I'd somehow known all my life that the disabled have to claim their right to life hand over hand.

After Roger died, his father admitted to me that once his son went blind, even with the reclaimed vision, Al figured the battle was over. He couldn't stand to see Roger hobbled like that and came just short of feeling his son was better off dead. I disagreed mightily when he said so and cited my brother's life of courage, then cited Roger himself. Because I know how unflinchingly he rose to the challenge, all through the summer. In his place I would have been long gone.

So it wasn't an accident that we began going out again to restaurants in the evening. Patched together with AZT and a cone of tunnel vision, instructed how to steer through public places, we knew it was now or never. We'd only go to places that were familiar—Chinese, burgers, Sunday pancakes at Pennyfeathers—but that was exactly what we longed to get back, the quotidian occasions of the neighborhood. Whenever we'd visited Madeleine in Paris she always took us the first night to her favorite *restaurant du quartier*, which no outsider could possibly know about. It's a stretch, I know, to see Hamburger Hamlet on the Strip in quite that way, but it's how it felt to us after eight months exile. And once we had broken the drought with tentative forays of our own, we began going out with friends again.

There were symptoms to deal with too, of course, despite the fort of normalcy we built against the blindness. Roger began to have fevers and sweats, one of the nonspecific gray areas of the disease that can be quite debilitating all on their own, frightening because they could be the beginning of some demonic infection or nothing at all. In the beginning Roger was more assaulted by fevers than by sweats. The zigzag graph of high temps and drenching sweats would start in earnest soon enough. Tylenol was sufficient for the time being to bring the fever down, but often Roger would wilt with a temperature after lunch and have to spend the afternoon in bed. That was the worst aspect of it, that it laid him low and stole so much time.

One poignant day I remember, Roger had made an appointment to see an acquaintance about drawing up a will. Alexander was prominent in the L.A. art world, very cultivated and silver-tongued, and we weren't certain that he knew about Roger's situation. So Roger was a bit nervous that he'd neglected to mention his limited vision to Alex, or that he'd be taking notes with a tape recorder. It may even have been that phase of the AZT cycle where Alex would have had to wear a blue mask. I assured Roger the meeting would go fine, that I would be there to show Alex in and help explain the ground rules.

Roger actually put on a shirt and tie, and after lunch sat in his leather chair in the living room to await Alex's arrival at 2 P.M. And he waited and waited, till three or three-thirty at least, before he conceded that Alex wasn't coming at all. I think I was more furious than Rog, who shrugged it off and took a nap. It turned out Alex had called the office in Century City and left word that he had to cancel. The lawyers in the suite still allowed Roger to use the main number and took messages for him, so as to give him the professional edge of a number that ended in three zeros. But this time someone at the switchboard had slipped up.

The reason Alex had to cancel was that his friend Tom was dying in a hospital in Santa Monica, with what was first stonewalled as food poisoning. Later, I believe, it was leaked that Tom had been in India recently and must've picked up something awful—perhaps the same exotic jungle horror Cesar had once put faith in. Tom died about three weeks later. So I had to let my anger at Alex go, but I still can't stop seeing Roger sitting there dressed up and waiting, feverish and nodding off but determined to carry through.

Al and Bernice were out again for a week at the beginning of June, and thereafter they came about one week a month. The fevers had so far been manageable, but there came a day when we couldn't bring it down from 103. Roger was feeling miserable. It happened that Dennis, the attendant, was off for a couple of days, and his replacement was one Scott Brewer, a registered nurse from Florida working at APLA as an aide because he hadn't got his nursing credentials yet in California. Scott was very skilled at making Roger comfortable with the fever, but he also thought we should bring Roger in to be sure it wasn't anything serious. "In" was all one needed to say anymore to evoke the world of the tenth floor. Scott managed to calm us down by saying he was fairly sure the problem was dehydration, since it was difficult to keep replacing the fluids lost to the fever/sweat cycle. When Cope concurred that we should come on in to the emergency room, Scott went with us.

As we waited in a cubicle room for Cope, Roger asleep on the gurney, Scott told me his own lover had died a year before in Florida, barely in his early twenties. The diagnosis of PCP took over a week to pin down, so the drug that would've stopped it was given too late. Within days Scott lost his nursing job and was thrown out of his apartment. He came to California to depressurize with a friend, who two months later was down with AIDS himself; Scott stayed on to take care of him. Now he'd volunteered at APLA to work for a fraction of a proper nurse's wage because he believed he was necessary. He was.

Roger was admitted right away and given the whole battery of medieval tortures, from bone marrow to spinal tap. Then we began the long grim wait for results, which trickled in over the next couple of days. In fact Roger turned out to be seriously dehydrated and was put on an IV drip. But the symptoms of that dehydration—weariness and endless sleep, so little responsiveness the second day that I couldn't tell if he was disoriented or not—kept us all anxious and terrified. From me it demanded the same urgent coaxing and engagement I'd used during the days of the brain involvement.

As each test came back clear I was more and more certain we'd pull him through, even as the temperature would swing to 103 or

104, followed by a typhoon sweat. The nurse had to monitor him almost constantly, and for the first time we engaged a private duty nurse to stay with him during the night. We didn't want him lying there chill with sweat and no one to change him, waiting for a floor nurse to make her rounds at night. I didn't realize just how frightened Al and Bernice were, or how off the wall was Sheldon's Malibu perception of it all. I only heard it afterwards, but on that second day he drove in from the beach, cased the situation in Roger's room, then took the parents downstairs to the cafeteria and broke it to them. "This is the end," he said. "You have to prepare yourselves."

His worst-case prognostication didn't tumble out till later that night, when Cope was in and the crisis seemed to be turning. Roger was sleeping comfortably as Cope told the parents all the tests were negative. Rog was going to be all right. Al started to shake with relief and said: "We thought he was going today. Our hearts are breaking, Doctor. Not just for him but for this boy too." And he pointed at me across the bed. I demurred and assured him I was fine. All I cared about was that we'd come through. No need for any broken hearts.

I think Rog was in for five or six days that time. I'd stay till 1 A.M. and talk to him while I changed his sweat-soaked gown, sometimes every hour at night. With the fever broken, he'd feel more comfortable. Then he'd perk up to have me there, and we'd fall into our private ironic shorthand. That's why the night sweats never scared me much, because I could see how hard the body fought, kicking free of the viral quicksand. And when the sweating was done with, there he'd be. I remember bringing him home on a Saturday morning, up through the back gate into the garden. He was choked with pleasure to see it again, even partially and dimly. For a while all he wanted to do was sit there and absorb it, the breeze and the smell of gardenia, the dog lying beside him, paws over the lip of the pool.

Then began our best reprieve, most of the month of June. Of course we still had a fair amount of checking in to do. Twice a week we'd have to go to UCLA to have the blood drawn and visit the pharmacy.

Everyone in the clinics knew Roger and me: we'd been in and out of there for a year and a half. And these people have a real connoisseur's appreciation of the fine points of war and the way men fight it. Even on bad days we tried to be up for them, the receptionists and technicians, for their morale was as much at stake as ours, and we had to help each other. The man who drew the blood at noon was Tonio, an unruffled Filipino who was as plainspoken as Rog. When he'd lean over to prick Roger's skin, they would talk quietly to each other, close as brother monks. Tonio was gay, as was a goodly percent of the hospital staff. Since the plague they had been laboring under an extra load of burnout, though their sympathy and compassion never seemed to fail.

Then once a week we'd have to go see Kreiger. The infection stayed in check, and the millimeters held. Kreiger was pleased by the progress of Roger's vision, though it wasn't ever fast enough for us. Dr. Martin came to the house on Wednesday afternoons for the fifty-minute hour. Meanwhile Dennis the attendant was a necessary figure in Roger's everyday life now. He'd help Rog get up and dressed in the morning, serve him breakfast, massage his legs, sit with him by the phone dialing Roger's business calls. Dennis made it possible for Roger to gather himself and spend his liveliest hours in full command of his battered resources.

And thus we had our own good time together—quality time, we call it in the mortal department, just as in the parental department. Always at midday for an hour, once I got going and before I'd meet with Alfred. Then again in the late afternoon, reading or out for a walk in the canyon. Then dinner and afterwards calling around to the family, and late at night the islands of time at 1 or 2 A.M., when he'd wake up and want to talk. Once we even called Jerusalem in the middle of the night, to wish Rita happy birthday.

Besides which it was summer, and Roger got back in the pool again. At four or four-thirty, with the white sun streaking through the elm trees, he'd do maybe fifteen or twenty laps. Especially if the two of us were swimming at the same time, we were suspended from all the misery, twinned and afloat as we'd been in the dolphin blue of the Aegean. Quality indeed. It must have been around then that Roger said, with a pained wistfulness, "If only it could stay like this for a while." A while is the kind of modest goal you spend your life searching for.

After midnight, during the hours when I used to sit and work, I'd be cleaning drawers and closets, tossing out masses of irrelevant clutter. When I worriedly complained to Sam that I felt as if I were throwing away the remains of people who'd died, he said it was entirely appropriate to clean out all the excess in one's fortieth year. The more I tossed, the more I felt I was following Thoreau's triple command: *simplify, simplify, simplify*. I remember going through drawers in the bathroom and finding Roger's contact lenses in their case. I realized he wouldn't ever be wearing them again, but was afraid to throw them away too, lest I discard the hope that held his vision. Two or three days later I finally steeled myself and stuffed the lens case in the trash, but guiltily, mentioning it to no one. And once I'd got rid of the lenses I combed the bathroom for every bottle of lens solution and all the eye paraphernalia that used to be so casually a part of Roger's kit. I also recalled a moment from ten years before: finding a card in his wallet not long after we met, which said, "In case of accident I am wearing contact lenses." Even back then I'd started to weep with dread, when nothing at all ever went wrong.

The closed-circuit televisions of the kind we'd seen at the Center were a couple of thousand dollars. Sometimes one would come in secondhand, but there was a long waiting list for these. By now Roger had come to grips with and compensated for much of the narrow bound of his vision, but the business of being unable to read was terribly galling. We were still waffling about investing in a TV of our own when Roger had a call one afternoon from Susan Kirkpatrick, an old friend from Comp Lit days who taught at UC San Diego. By coincidence, a great-aunt of Susan's, recently deceased, had used exactly the kind of unit we needed, and it was gathering dust in Susan's attic. Her husband was on his way up to L.A. for a biology conference the next week, so he would drop it off.

We set it up on a table in the brightest corner of the living room and began to play with its knobs and dials. I was so stupid about the closed circuit that the first few times I switched it on I tried to turn the volume up, when all it was designed to do was stare at a page and magnify it. Unfortunately, Roger was the only one in the household who could have made sense of the thing, but all he could do was squint at the blurred and tilted picture and tell us it wasn't coming through. We finally got it centered and focused right so he could read individual words, yet I remember countless occasions when he'd sit down and struggle unsuccessfully to make it render whole sentences. There was something wrong with the contrast, and the periphery of the screen was blank. I don't know why it took us so long—I only know I feel guilty about it—but it wasn't till late in the summer that we finally got hold of the proper serviceman. And by the time it was fixed Roger was gone, so all it ever really did for us was stand as a symbol of what might yet be given back, just slightly out of reach. After Roger died I arranged to have it donated to the Center in his and Susan's names, because I knew about that waiting list. "This will mean that someone can finish school," I remember Joey telling me.

June was rife with visitors from out of town, and if they were coming to say good-bye they kept it to themselves. To us it was all serendipitous. Richard Howard and his friend David Alexander, a painter, came out from New York on the way to comfort a friend in San Francisco, who'd lost his lover after a long fight. Richard read aloud to Rog a new poem, as well as a witty essay on baldness and a graceful obit for Jean Genet. We spent two lively evenings talking, and Richard was especially eloquent about Susan Sontag's *Illness as Metaphor*—a bracing caution about the scapegoating and self-blame that attach to certain diseases. We were all being assaulted now with the verbiage of self-help guerrillas who said gay men had brought AIDS on themselves. "I'm taking a course in miracles," as one West Hollywood airhead shared with me on the phone one night. "People pick their own diseases," he said, bragging that his lesions had faded to inconsequence.

Sally Jackson, a woman I once roomed with in Cambridge, was in town on business and called out of the blue. "So how are you guys?" she asked enthusiastically, and then sat silent while I told her the whole terrible story. She was one of those people back east who hadn't heard Roger was sick, and now she came by and made us laugh, leaving in her wake volumes of material on imaging and healing, and orders to eat brown rice. None of which managed to annoy me, because she was so dear. Perhaps, when it comes to the self-help business, it's all a matter of the source. We'd had no problem learning imagery from Rita, and in fact we did eat more brown rice as the summer lengthened. But nobody picks his own disease—except, perhaps, the more rabid religions.

I also remember Sally telling me over lunch that *I* wasn't going to get sick. Usually this bit of cold comfort made me quiver with rage, but I could see how she longed to make it all better somehow. The more I heard it the more I understood it as a need people had to believe the disease would stop somewhere—to save me if they couldn't save Roger. I try not to be offended by it anymore, and some dark side of me that lives under a rock presumably hungers for the assurance. Mostly it seems a necessary lie people tell so they won't go mad from the horrors of war.

The visit Rog took greatest joy in during the good month was from Peter Metcalf, an old buddy from Harvard who taught anthropology at UVa. Peter was born Cockney, grew up in New Zealand, and lived in Borneo once for a couple of years to study a Stone Age tribe. He arrived in L.A. when Roger was feeling most energized, when his vision had clarified to the highest degree it reached after the operation—maybe thirty percent of the left eye. Peter was an inexhaustibly antic man—"The only thing I ever wanted to grow up and be was a pirate"—and he and Roger reveled in old jokes and caricatures. Because he was also handy, we steered Peter around to various things that were falling apart in the house, which he fixed with dispatch. A girlfriend of his who taught anthro at an Ivy League school had told Peter that eight members of her department, grad students and faculty both, had AIDS.

Peter was troubled to see the goldfish struggling for breath in his spherical bowl, and he announced that we must go get Schwartz a proper circulation pump. For some time I had been operating on the theory that Schwartz had to fend for himself. I cleaned his bowl once a week and fed him his dead flies, but that was about as far as I'd go. So it was truly an otherworldly errand to go to the tropical-fish mart, whence Schwartz himself had come, for a pump and a bigger tank. Peter busily set up the new pet exhibit, and now Schwartz swam around the tank with delirious energy. Within a week his gold had come back shiny again. This mattered because Roger could see Schwartz through most of the summer if he peered close to the glass.

Peter also went with me to buy big-watted bulbs, which we screwed in all the lamps and overhead sockets where Roger sat at night. But the moment that clutches at me still happened the next afternoon, when the three of us drove up to Laurel Canyon Park to run the dog. There had been a great storm of protest at the park over the issue of leashing dogs. A few years earlier the dog people had reclaimed the park from bikers and druggies, so they figured their dogs had dibs. The county disagreed, and you couldn't go to the park without confronting a barrage of canine agitprop and petitions to sign.

We'd been telling the park saga to Peter, and when we got there he took Roger's arm as we headed out onto the knoll. There was a sudden ruckus of leashless dogs, and Peter turned to look at them, for a moment letting go of Roger. Peter and I were watching the dogs, and Roger walked into a tree branch, which poked his forehead—didn't break the skin, but he cried out, startled. Peter and I spun around in dismay, to see Rog clutching his head. A couple of people who didn't understand there was blindness here started laughing. The sharp end of the branch had poked not two inches above the eye we'd been fighting three months to save. From then on I would tell myself over and over, whenever I walked with Rog outside, to stay alert.

On Monday night, as Peter and I were leaving for the airport, he went in and gave Roger a last hug. Roger cried for a second, then grinned in a puckish way. "I promise not to fall off my perch," he said.

If constant vigilance about food had become second nature to us, we had to be equally alert now about fluids. The sweats and fevers waxed and waned through the summer, but even so Roger had to get into the habit of drinking a glass of water every hour, or lemonade or Ensure, a potent nutritional supplement. If nothing else, he had to counteract the effect of all that medication on his kidneys. It can be a wearisome business when it seems every swallow of water is another kind of medicine, with its own rigorous schedule. I can see how people debilitated by AIDS let either the food or the fluids go, it all becomes such a chore. That is, it requires a team effort. The side effect of so much water was that he tended to have to piss all the time. He began to keep a plastic urinal bottle beside the bed, so he wouldn't always have to be getting up at night.

Did all this mean he was more of an invalid now? I don't think that's how it felt to either of us. It was too peaceful being at home. We were so grateful having our time together, plus the excursions and visitors, that the summer bore the character of a recuperation, like taking a rest cure. Objectively, of course, the narrowing of scope and a life that was mostly bounded by the house were skating nearer and nearer to the thin ice of a hospice. But that is to forget how much we had been through—upheaval and exile and suffering—and what a luxury it was to do nothing much at all. I remember several evenings in June, lying in bed beside Rog and reading from the *Duino Elegies* of Rilke, and the two of us sighing with rapture over the waves of feeling. Another time I was moved by a profile I'd read in *The New Yorker* about Bishop Moore, and read to Roger a passage from a memoir Moore had written about coming home from war:

> *A man who has been to war will never be the same, for he has lost the virginity of living around the edges of life, and in the long gray waste of combat has had crushed his belief in smaller things. . . . It is enough for a man to live cleanly and quietly in peace with a few friends.*

That is what a whole generation of gay men are doing, as they care for each other and bury each other and take what respite they can in between. It all depends how close it has touched you, of course, how much you then feel that your near relations are all you really have. Only in the most extreme cases are people cashing out to go retire in a peaceful place, but I know four who have. Fast-track careers and the powers of money are the first to go, once you have been in the war.

John Orders used to tell us about his friend Lee, who died a few years ago in his late sixties—Methuselah time to us. Lee had survived his lover by a few sad years, and said to John once, "All that will matter to you when you're old is how much you've loved." That is as true of sick as old. When the summer was full, it felt as if we became a peaceful place for our friends to visit. Roger and I always used to talk about getting to a point where we could take whole seasons off—a farmhouse in Tuscany or Provence, with a string of invitations issued to all our friends so they would stream through and taste the pure empyrean with us. We'd done that once in Big Sur, rented a house by the crashing surf and shuttled five or six friends to visit. Since we wouldn't be doing the Provençal spring or the Tuscan autumn, we made do with the summer of Kings Road, for a while anyway.

And if Rog was something of an invalid now, held down by his lower energy and the shadow tunnel of his one good eye, all of that could be wiped aside by the briefest glimpse of what remained. We were walking one bright noontime on Harold Way, about fifty feet short of the Liberace gates, when Roger turned and peered at me. "I see you," he said softly, his lips curling in a smile. I laughed with delight, and we climbed to the top of the hill as if we could see to Africa.

Roger's parents were amazed and delighted at how well he looked—how well he saw—when they came at the end of June. It was probably the moment of peak efficiency for our various coping systems. Roger's law assistant, Stan, was coming a couple of afternoons a week, and though there was some distress over the dyslexic tendencies of the letters he typed, he was keeping Roger current on five or six ongoing legal matters. We were entertaining friends practically every day—I mean visiting with them, not feeding them as we used to. But there was a casualness and ease about Roger as he visited with people now. The first night Al and Bernice were in town we had them over for chicken, the four of us around the table, bright light above it, no masks and no one in bed.

They took him one day on an errand that would have had me bouncing off the walls, to the Social Security Administration to apply for disability benefits, amounting to two hundred dollars a week or so, which APLA had paved the way for bureaucratically. That night Roger was feeling well enough to go out for dinner, and we took the parents to Musso & Frank's in Hollywood because they loved its downtown funk. We were seated in the paneled red-leather bar, and though we would have preferred a booth, it happened that Sean Penn and Madonna were sitting at the next table in jeans and black leather jackets. I told all the Horwitzes to take a discreet look. I remember Roger squinting to take them in, and then saying, "Who are Sean Penn and Madonna?" Al and Bernice didn't know either, and it waited till they told six-year-old Lisa when they got back to Chicago before anyone got excited about it. But then Roger was always blissfully unaware of pop stars—the whole phenomenon went right by him. In '75, when we first visited Sheldon, he took us through Bel-Air and slowed at a pair of gates imposing as Blenheim. "Cher's house," he said. And Roger turned and whispered to me, "Who's Cher?"

It must have been during the June visit that Roger and his father were sitting alone in the living room one afternoon, and Al asked Roger what he would like to do "if something happens." Would he want to come back to Chicago? No, said Rog, he'd like to stay in California: "I've had so many happy years here." Is nine so many? I'd said to him maybe half a dozen times in the last twelve years, "I want us to be buried together, Rog. I don't want to go back to Massachusetts." I don't recall if he shrugged or nodded or answered me, but whatever it was, it was low-key affirmation, as if it wasn't an issue that mattered a lot to him. About a month after that exchange with his father—unknown to me—I stumbled out something about funeral arrangements. We were stopped at the light at La Cienega and Santa Monica, and I said we'd never really talked about what either of us wanted at the end. Roger pulled back with a certain distaste and said, "You take care of all that." From such fragments you have to make your way when the sky goes dark.

Were we sad? Not after a month of respite. We were just going along. I recall watching the hoopla that attended Liberty Weekend, and going in and giving Roger updates from ABC. We reminisced about '75 and '76, the Fourth parties on the gravel-and-tar roof at 142 Chestnut, looking down on the Charles and the Hatch Shell, where the Boston Pops held forth to half a million people. We hadn't required the television then to tell us how fine the light show was. On the fifth of July in '76, Cesar had departed for California to start all over. Before he left for the airport he scribbled a note in my journal, thanking me for the party and the "happy ending" to his years in New England:

*If later on, as we read this, we might think "How happy we were then!" at least we'll have that. That as we lived them, these moments, we knew they were important, and that's all there is.*

Ten years later, Roger and I were sitting on the front terrace having knockwurst and baked beans for supper, the summer light fading slowly as neighborhood firecrackers bulleted the canyon and revelers started to honk below on Sunset. Roger looked out through the coral tree—no, not so much looked as turned to the breeze—and he said, "We're living on borrowed time, aren't we?"

"Yes. Except lately we seem to be borrowing an awful lot."

It was only a few days later that Kreiger flashed his glass at Roger's eye and announced with calm dismay that the infection appeared to be moving again. Had I neglected to hold my breath when he made the examination? He told us we'd have to increase the acyclovir considerably, and the mode of delivery would have to be by IV. We were too aware of the need for immediate action to worry about the implications or the logistics. Cope and Kreiger quickly conferred, and they told us the IV could be administered at home by visiting nurses. They would help us make arrangements through a service for someone to come three times a day. There was some hope the course of the medication would only be a week. I remember, driving home from Kreiger's, that the most Roger could muster was a weary shake of his head and a mordant "Oh, God."

But we quickly got used to the schedule because we had to, and Roger had extraordinarily charmed relations with the group of nurses. We liked their style, these women who moonlighted at home a couple of days a week and let their hair down and dished every hospital in the city. Roger of course was the one who had to go through the needle sticks, as the IV apparatus was changed from vein to vein every few days, left arm to right arm. He would purse his lips in a whistle and suck in breath when the needle went in, and didn't complain or dwell on it.

But after about a week of treatment, the condition of his veins was a real problem. Acyclovir was very caustic, and the veins in Roger's arms were already shot from so much IV. The vein would "blow" sometimes after only a couple of doses, and then a nurse would have to search for another, sometimes even dig for one. For a while at least, we were able to count on the skill of the best of these women, who could slip a needle right in the first time. But the nurses themselves began to raise the issue of Roger's getting a catheter implanted in his chest, where direct delivery of the medication into the artery precluded the problems of the collapsed peripheral veins. He'd find it so much easier, they said, and who knew what other medication he might need in the future?

We were not big fans of future talk. Roger was adamant: Under no circumstances would he go through catheter surgery unless it was absolutely necessary. If the acyclovir was going to be short-term, he would steel himself to the needle sticks. Compared to the catheter's invasion close to the heart, the discomfort of the bee stings was tolerable to him, though I would flinch if it took more than one stick to find a vein. The problem was, we saw Kreiger every week, then twice a week, and he kept saying the same thing. The infection was holding back again, but he didn't dare withdraw the acyclovir. We tried not to plead about the trouble, expense and pain of the IV, and Kreiger certainly didn't advocate the Hickman catheter implant, but he couldn't in conscience stop the drug.

The vein situation grew more intractable as the days passed, especially when a new nurse came on, or one who had a lousy needle technique. One Saturday midnight the weekend nurse was clearly freaked about AIDS. She missed the vein a half-dozen times, till Roger was unhinged from the stress and discomfort. I pleaded with her to call someone else, but she finally slipped in a tiny butterfly needle, something she'd use for a baby. "This man don't have any veins," she said as Roger dozed through the IV, exhausted from the trauma.

Then we found out we'd used up all the nursing privileges covered by Roger's insurance, and we'd have to start paying for the nurses ourselves. I think it amounted to a thousand or fifteen hundred a week, and now we seemed to be more and more assaulted by the system we'd bought into, and yet there was no way out of it. Sometimes there was a glitch, and nobody showed up. Then I'd be frantically on the phone, while various answering service drones informed me it wasn't their problem, and why didn't we go to the emergency room.

One mid-July night we had to. Cajoling and whining at the service did no good, and we were afraid if we missed a dose it might turn out to be the final branch that poked the eye. So we went to UCLA at 1 A.M. and didn't get home till after five. And when Roger was finally sleeping in the cubicle at Emergency, the drug going in his arm, I took a walk around the campus—brisk and intense as the walk I took the day of the eye surgery, the same route even. But this time feeling utterly numb, just trying to get through the one night and sweat off some of the pounding anxiety.

It's hard to describe the strange double nature of the house in July, for if all this arduous drama over the IV was constant, three times a day for an hour, we also put it aside in between, or what was the point of being home? The friends still came, we still stepped out for dinner. We went after the windows of time with renewed intensity, seeing by whatever light was left. Alas, Roger was realizing with a growing sense of fatalism that his sight wasn't going to come back any more than it had, which wasn't much. Not quite shadows, not quite blurred, but only vision enough to see at home, where everything was the same, where the furniture is so actual it furnishes one's dreams.

We'd always have to be back by ten-thirty or eleven, when the night nurse would be coming by, so there was a Cinderella clock ticking behind every foray now. One night Kathy Hendrix had us to dinner with four or five friends, but it was difficult for Rog to separate so many voices, and his isolation tired him. After dinner Charlie passed around pictures of his new puppy. When they came to me I described each one to Rog and then passed it on, till everyone was sad and introspective, full of hollow dog remarks but thinking about blindness. We got home a bit late, or the nurse was early, and she had a tantrum at having been kept waiting. Roger shook off the Nurse Ratchit reprimand, but I was pissed, sick of the tyranny of their system. Besides, they couldn't get the goddam drug in his veins. Yet I made no recommendation for the catheter implant, because I kept praying the need for the drug would be over soon.

On Sunday, July 20, there was a black-tie dinner at Sheldon's house for all those who'd sponsored tables at the last Gay Center dinner. It was a thank-you party designed to get the ball rolling for the next fund-raiser. During the week or so before the twentieth we kept shying away from saying we'd come. The IV business was too intense, and we had no idea how Rog would feel. But Sheldon gave us the leeway to decide at the last minute, and when it turned out to be a beautiful day we decided we were in the mood for a midsummer jaunt. So I hauled out the tuxes. Roger couldn't fit into his pants—his weight was steady at 147/148, and he had a belly now from all the rich food. So he had to wear my pants, and I was able to suck in my gut enough to wear his. Since his tux has a satin finish and mine is a wool worsted, we were very Mutt and Jeff in the fashion department. Yet there was something so giddy and promlike about the dressing up, even the running around passing the pants back and forth, that we drove to Bel-Air in a jocular mood.

Still, it was a wearing business dealing with the crowd and the whole choreography of the event. We walked across the living room, Roger on my arm. I nodded to various people, and Roger didn't, of course. Our brothers looked over at us soberly, but stunned and frightened too: *There but for fortune* . . . We made it out to the terrace; the dinner would be served there and the light was best. We stood by the wall where it looks out over an unbelievable view at sunset, Mount Baldy to Catalina. Roger smiled at me shyly. "Well, here we are." Then Sheldon's current friend came over and gave us an ostentatious greeting, with a stage kiss to Roger's cheek that made me want to claw his eyes out, I was so afraid of the germs.

Friends appeared from every side to shore us up, however, and once we were seated for dinner the strangeness dissipated. Rand had arranged the seating carefully, so the people on either side of us were at ease as I told Roger where everything was on his candlelit plate. A full moon rose in the middle of dinner as if on cue, so luminous and close that even Rog could make it out against the dark of the summer sky. Rand and Eric Rofes, director of the Center, gave sober pep talks on the symbolic importance of the Center in a time of tragedy and backlash. The Burger Court had just announced its cruel ruling in *Bowers* v. *Hardwick*, in which the sodomy laws in Georgia were allowed to stand and no inherent right of privacy for homosexuals existed. The Constitution plainly did not include us. For days Roger had been talking passionately about the case, praising the stirring dissent of Justice Blackmun, and impatient to hear every word the papers printed on the subject.

When Rand spoke about AIDS I squeezed Roger's hand under the table and happened to look over to the next table, where I caught the eye of a man who'd just been diagnosed with KS. We exchanged an ironic smile. When the full-moon dinner was over, Roger and I had to hurry to get home in time for the evening dose. Sheldon came over and chatted a bit before we left, awkwardly filling up space with banter about his date for the evening. We greeted him pleasantly enough, though by now he was high on my shit list. He'd told his father two months before, after the dehydration scare, that he'd check in and visit with Roger three times a week. I don't know where he came up with that symmetrical number, but in any case he hadn't been up to visit at all. Tonight was the last time but one that Rog ever saw him, however dimly. Yet the black-tie affair at Sheldon's was a true high point of the summer, even if I had to put away our tuxes, with their proper pants, wondering if we'd still be around for the Center dinner in November.

Just after we got back from Greece in '84 I'd bought a book of photographs by Eliot Porter, *The Greek World*. Porter had taken the first picture Roger and I ever bought together. His Greece book was full of magnificent light and eloquent shots of every totem shrine, but when I picked it up now, two years later, I began to read the text, an evocative historical essay by Peter Levi. I'd rattle off bits of it to Roger every few days, and sometimes a passage would send us back to the guidebooks and postcards. Then we'd talk again about the early morning in Delphi or the walk through Phaestos in Crete, the whole ruined palace all to ourselves.

From my reading in fits and starts I kept coming across references to Socrates being put to death in Athens, and commentators would refer to the event as if one already knew the whole story. But I didn't. I vaguely remembered from freshman philosophy at Yale that Socrates was thought to be some kind of heretic and was tried and executed for having a bad influence on the youth of Athens. Yet it made no sense to me now that a state as perfect as fifth-century Athens should kill off its wisest citizen. So I told Roger perhaps it was time we read some Plato.

One evening during the IV struggle, Kathy Hendrix dropped by after supper with Jill Halverson, for whom Roger had done the legal work for the Downtown Women's Center residence. Jill gave Roger an update on the shakedown period that followed the May opening of the residence, designed for homeless women disabled by chronic mental illness. At first some of the women had slept on the floor instead of the beds because they were used to bedding down on the street. During one of his hospitalizations a card had arrived for Roger at UCLA, signed by all of Jill's ladies. When she received a social service award in 1987, Jill accepted it *on behalf of the women of Skid Row and in memory of Roger Horwitz, who even while he was dying of AIDS volunteered his legal services so that women without a home might have one.*

Kathy told a story that night about a friend of hers at the *Times* who'd gone to Cabo San Lucas for her honeymoon, staying in an old hotel that was built along a cliff face hollowed by caves. As she and her husband slept, a creature that turned out to be a sort of equatorial raccoon crept through a crevice into their room. Kathy told the story in wonderful slapstick detail—"Honey, where's all the fruit that was in this bowl?" In the middle of the night they woke up to find the creature curled at the foot of the bed, asleep. They about died. Roger grinned expectantly throughout the telling, then at the end started to laugh—but I mean he laughed till he gasped and the tears streamed down his cheeks. He rocked in his chair for a full minute, holding his side where the shingles were. For a moment I was worried he might hurt himself from the strain, but it's hard to stop a man who's been through so much pain from laughing. Days later he'd suddenly think about the raccoon on the bed and start whinnying with laughter again.

Still he was on and off AZT, his white count followed closely so there was never a repeat of the tenth-floor isolation. The calls and inquiries about the drug had never ceased, and now they only intensified as it was bruited that the FDA would soon relax controls and permit wider testing. All the other antivirals were a bust next to AZT. There was hope that if the drug was given soon enough, before breakthrough to full-blown symptoms, the side effects to the blood would not be so severe. Ten men I know are on it now and cruising along just fine, no breakthrough at all. The rage is still unquenchable when I think if the drug had been on line a year earlier, then Roger would be cruising along as well, even now. The "what ifs" do not go away. Of course I knew to be grateful that he got AZT at all, just as he and I both were certain it brought him back and was giving him time. But the longer I watch the government do nothing, the months thrown away with the lives of my friends, the more I see it didn't have to happen. The drug was there on the shelf in '83 when they finally pinned the virus down, but nobody bothered.

We'd reached the end of the line on the IV issue. Gently but insistently Cope and Kreiger both came down on the side of the catheter, and finally Roger couldn't take the needle sticks anymore. Everyone was united in saying he'd have no problem with it—there were cancer patients who'd had catheters in for three and four years. We'd be glad of the independence, they said, and now there was no reason we couldn't take over the IV from the nurses. It all made sense, and I was convinced. It was only when we had to make the appointment for the "procedure" that Roger admitted to me why he'd been reluctant for so long.

"It's the beginning of the end. I know," he said with a certain anguished resignation. I recall being startled and leaping to reassure him. Again I wish we could've had the talk about how it felt, the dread of the end beginning, but something blocked me from seeing it that way. I believed in the catheter the way I believed in AZT, with missionary zeal, if only because the needle pain was going to stop at last. Roger must have seen the whole intrusive procedure as a moon appendage that would never go away, but that was the only time he ever said as much.

It was scheduled for Friday, August 1, and happily Jaimee and Michael decided to fly in from Chicago for a couple of days before. If Roger wasn't quite so perky as he was when his parents visited in June, he still looked terrific—no, that's an exaggeration. But he was up to his weight; his color was good; he was mobile and alert. Besides which, he was proud of having legal work to do every day, and he spoke with Jaimee and Michael as if there was all manner of things to talk about besides AIDS.

While he was napping Jaimee went out and bought him a yellow sweater. I'm sure he was able to see that color when he opened the package, unless it's just the joy of the moment I'm remembering. Maybe we told him the color. Sheldon drove over to say hello to his sister but begged off going to dinner, finally coming to the City Restaurant for coffee because Michael insisted. The dinner was clumsy, with Sheldon talking about the bank and never addressing a remark to Roger, and Jaimee and me on either side of Rog, helping him through the intricacies of the meal.

Families do not always come together neatly in a tragedy. By now Roger and I had given up on getting any emotional support from Sheldon, but that night for the first time I realized how gray and frail Sheldon was looking. He'd always possessed an exuberant vitality, especially on the playing fields of power. Now in his mid-fifties, twelve years older than Roger, he seemed like an old man as he sipped his decaf. Was he sick? Was he *pre*? I didn't even want to think about it. But I only grew more suspicious when, a few weeks later, he went away to a health spa for a couple of weeks of rest. He wasn't the resting type, and all the while he was gone he left no number where we could reach him. He would call in, cheerful and upbeat, but the calls were always brief.

On Friday morning Jaimee and Michael took Roger over to UCLA to check in, but they had to leave for the airport at noon, and the surgery was late. We had a meeting with Dr. Ahn, the surgeon, who apologized for the tardiness. It was difficult on a Friday to get the one operating room that was fitted out for isolation. I don't remember much about that afternoon, keep mixing it up with the duplicate operation six weeks later. I know it didn't take long, and remember we had a quiet evening on the tenth floor after we called the families to say it had all gone swimmingly.

The catheter was a length of flexible white tubing, about eighteen inches long, with a two-pronged end so there were two ports for insertion of a needle. Between doses the tubing was coiled and taped to his skin, beside the dressing that covered the tube's entry into the chest, a nest of stitches around it. Charlene, one of our regular nurses, showed me the elaborate protocol for cleaning a port before inserting a needle. Six swabs with alcohol, six with Betadine, using a circular motion with each swab. I watched her do it and realized I had to learn it all, and it seemed ten times as complicated as the eye medicine, the chance of infection so close to the heart frightening.

Roger's demeanor was pretty flat that night, and Charlene observed laconically, "Roger, this is the most wimbly I've ever seen you." I registered the solecism—wimbly—with all the dispassion of a writer, Bill Safire jotting a note for his language column. That's when she told him, "Hey, you're the miracle man," as if to say he had to keep going. When the tenth floor grew quiet, I massaged Rog and watched a documentary about Winnie Mandela on PBS. For a whole hour I sat riveted, my eyes smarting with tears as she spoke of her children being taken away from her, the years and years of separation from her husband, her growing politicization. They take your life away whether you fight or not, so you might as well fight. Fuck *Bowers* v. *Hardwick*.

Roger was in just overnight, and we came home and moved to restore the island of the summer. The wimbly feeling of frailty passed soon enough, though Roger never quite got over his squeamishness about the catheter. I'd tell him the whole thing would seem a lot less strange if he could just see how it worked. Somehow the mechanics of the system, watching it function like clockwork, got me past the alien part. I immediately began to study the nurses' technique, hopeful that I could take over at least one of the daily doses. The pool bedroom was piled waist-high with cartons of supplies and equipment, the IV pole beside the nightstand. IV procedure through a catheter involves dozens of steps and a lot of English, from mixing the medicine with sterile water to constructing the hypodermic to preparation of the port.

Then once the needle is in, you have to monitor the flow, a drip about every four seconds so the drug goes in over a full hour. The nurses were good enough at it to be able to do their paperwork while the drip went in, while I would usually watch it like a hawk for the whole hour. But I learned my lesson well, and by August 10 I was able to do the late-night dose myself, without supervision. Two days later I took over at 4 P.M. as well as midnight, so we only needed a nurse in the morning. The 8 A.M. nurse would change the dressing too—another three dozen steps of protocol, which I didn't want to get into—and I took care of the rest, thus reducing our nursing bill from nine hundred a week to fifteen hundred a month.

But our staffing changes were more extensive than just the nursing department. Stan was proving to be out of his depth when it came to the arcana of legal correspondence, and I found myself retyping his letters late at night, so Roger finally let him go. Happily, we had an old friend, Fred Sackett, who'd been trying to think of a way to help us. Fred was between executive jobs and contemplating a move back east. In the meantime he volunteered to work at the same slave wage as Stan, and early in August he began to come by twice a week. Fred was so vastly overqualified it was funny, but he made an enormous difference to the texture of Roger's life. They quickly sorted through great tangles of files and bank accounts that I just couldn't handle. Roger looked eagerly forward to spending time with Fred. They were about the same age, and Fred was from West Virginia, with a rolling drawl of humor dry as country dust. He kept Roger's irony up.

Then Dennis the attendant gave notice to APLA. He'd grown tired of the daily commute from Long Beach and wanted to work closer to home. I wasn't so sorry to see him go, but Roger was: "He's so sweet to me, Paul. He's so peaceful." It was our great good fortune that Scott, the widow nurse from Florida, was just coming off a case for APLA and free to start with Roger right away. That is, he had just buried the previous case. Scott was immensely helpful to me, because he had all the IV training and could spot me when I did the four o'clock procedure, though he legally couldn't touch the IV himself. He possessed an innately sunny temper and reservoirs of enthusiasm, despite the fact that he left us every night and went home to take care of his dying friend. I was glad of his volubility for Roger's sake, and just as glad to be rid of too much peacefulness from Dennis, fearing anything that smacked of disconnection from life.

Sometimes I would go into Roger's room and he'd announce plaintively that he couldn't see much at all. Then I'd notice his glasses sitting on the nightstand and tell him to put them on, and suddenly things would look better again, and we'd laugh. Other times there was no quick fix. It wasn't so much the shape and clarity of things he'd lost, but light itself. The part of the retina that saw light had diminished into dusk. Thus the most wrenching thing of all was to walk in after sundown and hear him say, "Are the lights on, Paul?" And of course they were. Maybe this part came later in August, yet it fits the particular twist of fate that took hold after the catheter surgery. Kreiger looked in Roger's eye and declared the infection hadn't moved, which was good, but now he could also see a cataract starting to form. So week by week Roger would begin to lose the precious lines on the "E" chart that he'd gained since the April operation.

Rand Schrader came over every Sunday morning all summer long to have breakfast with Rog. I'd leave the coffee ready to go the night before, and Rand would arrive with croissants and help Roger get up and dressed while I slept in. Early on, in May and June, Rog would sometimes say how he didn't want to die, but with a rueful unspoken acknowledgment that that's where things appeared to be headed. As time went on, the matter of death came up less and less, but who knows if that is the same thing as acceptance? With Roger increasingly compromised by his illness, Rand would have to search for common ground, since he didn't want to just chatter on about life out there, where the young were still young. So he found himself lobbing questions about Roger's years in Europe, what it was like to leave home at nineteen or to meet a man in Paris. "But they were nice and easy times," Rand insists, "they weren't the least bit sad.

And Roger was all there all summer, till the last Sunday. He never checked out of the world."

On Tuesday the twelfth I had a battery of test results come back from Dr. Scolaro. The good news was that my T-4 number was up from 430 in May to 480, a degree of difference that nearly anyone in the know would put down to lab variation. When you're on the receiving end of the numbers you tend to say it's lab variation if the number has gone down, and if it's gone up you feel you have started reversing the trend because of whatever magic you're practicing at the time. In my case it was yeast and soy and eight different vitamins. Plus I redoubled my relaxation exercises, which I did instead of a nap, sleep having taken the summer off. The bad news was that the viral culture indicated the AIDS virus was active in my system, not dormant and tucked in a deep genetic cave. Scolaro advised that I start ribavirin as soon as I could.

Mid-August was hot, and we usually wouldn't take a drive up to the park till nearly sunset. I remember walking with Roger across the lawn to a grove of beeches where picnic tables were set out in the shade. Though the light was bright gold and clear, Roger was seeing poorly; as I remember now, he saw worse when he had a fever and better when it was down. We sat on a picnic bench, and I read him the paper—unless by then we had started Plato, which I would read him in snatches several times a day, sometimes only a page or two. Today he wasn't listening much, and I could tell he was sad. So we just sat there for a bit, not talking, while I watched the various dogs and their owners cavort in the park. We used to call it *La Grande Jatte*, after the Seurat in the Art Institute.

Suddenly Roger began to recite Milton's sonnet on his blindness: " 'When I consider how my light is spent / Ere half my days, in this dark world and wide . . .' " I don't remember how far he got before he choked up and couldn't go on, but that didn't matter. Neither of us would have been very receptive to the bullshit about bearing God's "mild yoke." But I can't ever forget the moment, looking out at all the sunset yuppies and their dogs while Roger declaimed his loss in a broken voice.

Yet I also recall the Perloffs coming over to visit late one Saturday afternoon, while I was out doing errands. Roger heard the doorbell and felt his way through the house to the front hall, hollering that he was on his way. He opened the front door, and no one was there. "Marjorie? Joe?" he asked, and heard their muffled voices answering him from outside. He had opened the hall closet instead of the front door, just beside it. As he hastened to let them in, he announced, "I'm becoming Mr. Magoo." And told that story on himself, *laughing* at himself, for the rest of his days. That is the rhythm to try to understand about him, "On His Blindness" and Mr. Magoo.

The last photographs I have of him are mid-August, a hot Sunday afternoon in the courtyard under the carob tree. An old friend from Boston who was very big in public relations and worked like a team of sled dogs dropped by on the way to the hospital to visit a colleague down with PCP. I didn't discover the pictures till three months after Roger died, when I finished off the roll taking shots of Lawrence's grave in Taos. At first I was afraid to look close because they came so near to his death, but they're marvelous pictures, without a shadow. Peaceful, in fact, and sporting the sweetest smile and a dazzling white shirt as the sun plays in the courtyard. The right eye is still, because that's the blind one. The left eye peers and doesn't see much, but it sees me. In the country of the blind, as the French say, one-eyed men are kings. I am the one with the camera, who has taken his picture in every shady court from Athens to Kauai. The PR friend who visited that day was jovial and round, lamenting the lunch demands of his job, which kept him so fattened up. He was diagnosed with PCP himself five months later.

But August in my mind was mostly Plato. Those were the hours when we sat in the hurricane's eye together. We knew full well we had reached a summit, just as we'd known at Delos, facing the row of stone lions that guards the lake where Apollo was born. When you have the time to read a little Plato, when the other half of you wants to do it as much as you do, nobody wastes a moment worrying that he's wasting time. For the moment, no numb distractions are required, no maddening details to attend to. Jaimee sighed to Michael one night when he asked her how she was: "Fine. I watch the soaps all afternoon, and Paul reads Plato to Roger."

We used the blue-cloth Oxford volume called *Portrait of Socrates*, comprising the *Apology* and the two last dialogues of the philosopher's life, edited by the aforementioned Livingstone. Being good students, we started with the introduction, which took us nearly a week to read because we'd stop and talk about it so much. Livingstone's elegant sketch of Socrates' place in his own culture, and thus in ours, turned out to be a sort of lightning rod for us, who had profited all our lives from the gifts and whips of the intellect. Please, we were the only Harvard/Yale marriage on our block.

That Socrates was unpretentious almost to a fault, never believing wisdom was his alone or made him superior, but that everyone possessed it if one could only talk it out. That his method was conversation—this alone gave a sort of romantic burnish to the running Platonic dialogue the two of us had volleyed for twelve years. That he lived in a time of "official" gods, the golden age of his people seizing up as Athens fell under the wheels of war. It all made a certain Mediterranean sense, of connection if you will, which ought to temper my stridency toward those who contrive their essential self from the Bible, or indeed from Warren Beatty's sister. Livingstone imagined what a threat Socrates must have been to established ideas and self-important men. The more we thought about him, the more he seemed a first principle of all our own reading and travel and yearning for definition.

Then we turned to the *Apology*, Socrates' defense, delivered in the agora below the Temple of Hermes. Two years before, we had wandered the very spot, our guidebooks out like dousing rods, trying to make a city out of fallen stones. Socrates says—but you know all this part already, or at least you're supposed to. Or maybe you don't really need it till you're right at the edge of the cliff, as we were. "It's miraculous, isn't it?" a fellow poet, Sandy McClatchy, said to me on the phone one day when I was waxing on about Plato. There was nothing to figure out or understand at all. Just the clarity of it, unfiltered by vanity or bullshit or the need to kiss ass. And how tough Socrates is as he goads the pompous and self-deluded and dares them to put him to death.

"He sounds like Cesar," Roger would say after some effortless parry or thrust of rhetoric. And we laughed to remember our friend, the wagging finger in four languages as Cesar browbeat culture and self into his wayward students.

What direction were we going in? All downhill? It felt like a kind of stasis as the summer perceptibly ripened. Not that I have forgotten the times when Rog would cry out in pain, "Paul, I'm blind!" Or how, stirred by an aching nostalgia, he would shake his head and choke back tears and say, "I miss Cesar." But now was also the time when the drenching sweats would leave him awake and curiously refreshed at 1 A.M., and he'd ask for the tin of cashews and a glass of lemonade, and we'd talk and read aloud till three. I remember our friend Gordon came through from Canada and spent a couple of days doting on us, cooking gourmet dinners and reading to Rog, taking me to the beach for a swim. And that was all fine, but I could see how far gone Roger was in Gordon's eyes, in the year since they'd last been together. It only made me feel that our safest time was the two of us alone, when the shorthand kept us utterly in balance.

On Wednesday, August 20, Roger had blood transfused, about three units, as I recall. Though the white count is most consistently affected by AZT, the red count is also a problem. Many full-blown AIDS patients on long-term AZT have become transfusion-dependent. They have also gotten fairly blasé about the vampire part; but we still thought of transfusion as a grave and unsettling procedure. Yet the new blood perked Roger up considerably for several days, animated and energized him for work. He even talked with Esther Richmond about getting a new will written, the matter unbroached since we'd tucked the '80 version in a drawer.

He was coughing again, but even my radar wasn't especially flashing red. It was more than anything a clearing of the throat, and though in the past that very quality had been part of the ominous slippage toward PCP, I would not see it that way. Or perhaps I couldn't and still go on giving IV twice a day, not to mention keeping up food and fluids and twenty-two pills and Plato.

After one of his Wednesday sessions with Dr. Martin, Roger reported that he had admitted he'd had a good life. I remember how lucidly he repeated the phrase, amazed almost to be saying such a curiously final thing, and with no foreboding of death in the tone, or nothing gloomy at least. But I can't be sure, for I was behind in the death department. Even now, when I'm all caught up, it bewilders me to try to figure what he knew and I didn't. I had an appointment with Martin myself a month after Roger died. The first thing he said was: "He loved you greatly." Then he explained how Roger had gotten beyond the fear of death. Toward the end, he said, Roger's world was one of constricted hopes—Will I have enough energy to work with Fred? Will I be able to eat my supper? "It's impossible to conceive of ourselves without ego," Martin said, but that is where Roger arrived. I can't, of course, know what he must have thought in the hours when he lay there quietly, all but blind. I only know we never seemed any different, not between ourselves. And I felt no shadow of death when he said that his life had been good. Mine too, I thought with a pang of pride.

Near the end of August he spoke of being disoriented sometimes when he woke up, because he never knew in the darkness whether it was night or day. He spoke of this so precisely that I took it at face value. Martin prescribed the antidepressant Haldol, saying that many AIDS patients had found it helpful. Only he didn't just mean *blind* patients, did he? I never really picked up on the possibility that the disorientation might be organic. I didn't like Haldol and didn't see why it was needed on top of Xanax, but Roger stopped complaining about the problem, so I figured it must be working. Then one night at four or five, after I'd gone to bed, Roger got up and walked out the back door. He felt his way along the back of the house, skirting the lip of the pool and ending up by the pool equipment. When he finally came to himself and realized where he was, he called my name as he thrashed at the ivy. A neighbor shouted irritably from a window, "Go around to the door; you're in the bushes!"

I never heard Rog calling. Somehow he was able to retrace all his steps and come in and find me in the front bedroom. I shot awake and held him, scratched and wet from the bushes and so glad to be in my arms. He told me the whole story as I put him back to bed, and to me it was more of a dream disorientation, compounded into sleepwalking. If in fact it was dementia, it came on him for fractions of seconds only. The night by the ivy was the one extreme moment, but now I see it's like a dozen other stories I've heard of AIDS people wandering outside, very late on in the illness, sometimes into the ice and snow. With us it wasn't a pattern, and there was nothing to compare it to—except perhaps the lapse of speech in February, the so-called false alarm.

I cannot somehow put together those nightmare cries with the lazy Sundays talking with friends in the back garden, or the evenings reading Plato. Thus if things were darkening I did not see, and that is one of the cruelest ruses of the virus, letting you think the good times are the real times. Besides, I was locked in a true romantic's presumption now, with every page I turned. No matter how scorched the earth became, no one could take Plato away from us, not what we'd managed to read together. And I did think it consciously, even when he made the ironic comparison with Cesar, that it was Roger who was like Socrates.

He, of course, would have groaned with distress to hear such an outrageous exaggeration. Whenever I'd tell Cesar a Hollywood story thick with prices and salaries, Roger would always murmur, "Divide by three." But I don't especially mean that his mind was as fine as Socrates', or his integrity so unsullied. I only mean the honesty and simplicity, the instinct that he wasn't better or wiser than anyone else. I was too shy to say it out loud, but then, from here on, there was much that would have to go without saying. Whoever Socrates was, we read the blue book for the same reason, to see how a man of honor faces death without any lies.

# ·XII·

*8/25 Monday*

*V. difficult weekend, esp yesterday—we did not move from the house except a little walk up Harold Way. I lay on my bed at 4 PM & thought I'm just waiting to get sick.*

I kept telling Sam late at night that I wasn't exactly depressed, I was frantic, and I liked that better. I could neither hold to nor project a future anymore, and the consequent dread and rage had left me wildly manic. Sometimes I could feel my heart pounding as I counted out the day's pills from eight different vials, or ventriloquized a smile in order to talk business. Sam thought it was partly to do with the news I'd had about my active viral status. If I dared to slow down or think too much I'd end up looking blankly at the ceiling as I did on the twenty-fifth, staring into the coming storm.

Yet I don't recall that Roger and I were arguing or using each other as punching bags. We were much more drawn to comforting now, and curling like spoons to rest, talking softly of nothing much. If I suddenly panicked and told my fear of the calamity falling on me—the old horror of the two of us in separate hospital rooms, dying the same death—Roger would quickly force me to take a cold-eyed look at the reality: "You're fine." As if he would not countenance any moaning about an abstraction less savage than blindness. I am the same way now myself, ready to jump out of my skin if someone gets testy or whiny about anything less apocalyptic than AIDS.

Sheldon had been pushing for some time about selling the apartment house on Detroit, and now he grew insistent. I couldn't believe he was engineering yet another way of cutting himself off from us, and he seemed to have no feeling at all for how thrown we were by matters that smacked of final payments. His argument was that Roger and I could no longer keep up our end of the maintenance, and he simply wasn't of a mind to pick up the slack. I could see what a blow it was to Rog to lose the property he'd invested so much pride in. It all just seemed unnecessary, since Sheldon had a whole organization to manage his properties, but he apparently saw it merely as a business proposition that was going down the drain.

Roger didn't try to evade his nagging, but struggled to ask the right questions about the market, whether it was a good time to sell. After all that work he didn't want the investment to be a bust. He'd laid out five thousand once on a stock tip that went in the toilet, and he always worried afterwards that he might be dumb with money. I never indulged this brand of fretting. "Don't worry, darling," I used to tell him. "I'm a terrific investment."

Roger decided himself that we ought to get rid of the Datsun, since we didn't need two cars anymore and insurance in Hollywood was hopeless. I didn't really care, though I felt a certain sentiment for the gray 280Z, which we'd bought the month we arrived in California and which Roger had kept in mint condition. But a friend who worked at Sheldon's bank offered to try to sell it for Roger, so one afternoon I delivered it to her. I spent twenty minutes rooting under the seats and in the glove compartment, turning up ticket stubs and Stim-u-dents and quirky notes in Roger's hand, till I was overtaken with sobs. I had an irrational fear that if we gave up things like this, got rid of too much that bore his imprint, Roger would surely die. I was therefore limp with relief a few weeks later when Jennifer had to admit she'd had no offers, and we took the Datsun back. We stored it in the garage and cut its insurance to the bone. Six months later, when I finally got rid of the accursed Jaguar, I put the Datsun on the road with the strangest sense of joy, as if Roger were suddenly near again.

I scored my first batch of ribavirin that week from Jim Corty, an extraordinary hulk of a man whose passion for fighting fire with fire was as obsessive as mine. Jim was a nurse who cared exclusively for people with AIDS, and he personally drove a van over the border into Mexico every couple of weeks to haul back great quantities of ribavirin, supplying dozens on both coasts. His own lover, John, had been diagnosed in the spring, and Jim was constantly monitoring Roger's experience with AZT, eager to get it for John once the protocols were expanded. Jim always made me feel we would beat it, and never failed to rekindle my excitement about AZT. Ribavirin of course was a much less certain drug, but I went on it anyway, because there was no other game in town for me. I had been too vocal for too long that people ought to be getting tested so they could demand medication early, and it was time for me to put up or shut up.

I know I was growing increasingly desperate about Roger's cough, and if he suddenly had a jag I'd find myself getting irrationally angry, though I could usually swallow it. But I would have sworn there were no other ominous symptoms, no shortness of breath or overwhelming fatigue. This is not to say he didn't sleep a good deal, but between Scott and me we were very skilled at getting him up and going so he didn't sleep the day away. When he was up he was animated and alert, especially when anyone visited. The summer days were so lambent now, even as the summer waned—mornings in the garden while I read him the paper, evenings reading Plato, the smell of anise when we walked at night. These brief, immediate goals of the day-to-day we had come to cherish, no matter how constricted our movements.

It was Friday of Labor Day weekend when Scott asked me as he was leaving the house, "What does the doctor say the prognosis is?" I suppose I knew he was asking about the timetable of death, but that didn't seem to me the appropriate question at all. "The doctor says he's doing fine on the AZT," I replied, a bit defensively. Not that Cope had really said as much lately, but it was implicit in Roger's survival for nearly ten months now since he started the drug. He was the miracle man, period. He had to be, because thousands of our brothers were about to follow him on AZT.

A series had begun to run in the L.A. *Times*, a portrait by Marlene Cimons of an AIDS person in Boston. Jeff, his name was, and he'd been chosen to be in the AZT double-blind study being funded by NIH. It wasn't hard to get reanimated over AZT as news of its efficacy began to break in waves at last. We were thrilled by the *Times* story and very moved by the passion of the man's doctor, who reminded us in his patience and dogged persistence of Dennis Cope. So I tried not to overreact to the first bad news about AZT, which was Roy Cohn.

The press had uncovered the fact that Cohn was being treated at NIH in Washington, and the rumor was that it was AIDS, despite Cohn's drone of denial for the last two years. We had known through the grapevine for nearly a year that he was among the first AIDS people to go on AZT, after Nancy Reagan intervened in his behalf. The press was stumbling all over itself getting the story wrong about Cohn's demise, but I had a nearly day-by-day update from Craig, whose friend Donald was getting AZT intravenously on the same floor in Bethesda. "He's going to die in the next couple of days," said Craig, and I tried to keep the thought from racing in my mind: But no one's supposed to die on AZT.

CBS did a big report one evening that week about crack cocaine, the report we kept feeling they ought to be doing weekly about AIDS. Don't you understand? friends in New York would say, hoarse from screaming at the press for coverage. Cocaine wasn't a problem till it started turning up among the children of media dons and the Washington power elite. This at a time when I would hear at least every other week about the discreet death by AIDS of one or another rich man, the cause of death fudged on the certificate or otherwise unreported. Every gay man I know has stories of married bisexual men who died in the secret enclaves of family, town, church, and local GP, all without saying the "A" word. Even certain gay doctors, we heard, would blur a death certificate if the family was mortified enough.

Saturday before Labor Day we took the dog up to Laurel Canyon, and Rog was feeling well enough to walk all the way around the perimeter of the six acres, where it falls off steeply into bone-dry chaparral. Puck nosed among the trailside bushes, where the fleas are epidemic in the fall. Roger and I were arm in arm and slightly huddled, as we always were these days, but no one stared at us, wrapped up as they were in volleyball and holiday picnics. We'd reached the far edge of the park when a man with a pair of German shepherds came sauntering by, barking commands to his dogs. Puck preferred people to other canines and thus kept his distance, but on some mangy whim the two shepherds suddenly turned on Puck and grabbed him. I had to leave Roger to go hollering into their midst, and it almost came to a fistfight between me and the shepherds' Prussian master. Roger stayed calm, but I saw him straining to listen and separate the snarling of dogs and owners.

I couldn't let it escalate any further without causing Roger problems, so we left the field. I broke down crying in the car, overwhelmed by anything harsh or disruptive. Roger was very good about me crying now, where he used to get impatient and tell me to pull it together. He would let me weep it out, soothing me but offering no contradiction of the tragic. "My poor little friend," he'd say tenderly. "So many things to worry about. Come on, let's go home and have a spoon."

Sunday the thirty-first was our twelfth anniversary, and I decided to invite a few friends for dinner to celebrate. It turned out to be a bad idea, because I was beyond manic all day long as I did my IV chores and tended to Rog and tried to cobble together a proper cold supper. I would forget about people not knowing the rhythms of dealing with Roger's blindness, especially that it was easier for him to have one-on-one conversation. At the table the guests talked too much and didn't address themselves to Rog, so he ended up feeling ignored and lost. Meanwhile I was angry as I dished out the dinner. I couldn't wait for them to leave, and when they did I walked them downstairs, where they wished me a dubious happy anniversary, as if to say how could I be happy? "I'm just glad we're still here," I said evenly. "That's all I care about anymore."

Still we were going to UCLA three and four times a week, to have blood drawn and to see Kreiger. They were fairly dispiriting visits, and sometimes Roger would be so tired or feverish that we'd use a wheelchair to go from the parking lot to the clinics. Every appointment with Kreiger was sadder now, as Roger saw less and less of the chart, especially when he had fever. Kreiger avoided sounding falsely optimistic, but he swore they could do a quick laser surgery on the cataract, once it had ripened enough. Then Roger would have back the limited vision he had achieved in the summer. At least it was something to hold on to.

Wednesday, September 3, I had my first appointment with Dr. Wolfe, who'd left UCLA and was in private practice in Century City. I wanted his measured and cautious approach to treatment as a balance against the aggressive approach of Dr. Scolaro. Wolfe for instance didn't think much of ribavirin, I could tell. The visit was smooth till I was driving out of the parking lot and suddenly faced the twin office towers where Roger had had his digs, and I burst into tears at the sight of what he'd lost. I wouldn't go to Century City for months thereafter.

I came home to find the dog had a running open wound on his leg, and we realized he'd taken a tooth tear from one of the German shepherds. So I made an appointment with the vet for Friday morning, which is why I didn't take Roger over to UCLA for his blood transfusion. Cope called Thursday and asked him to come in next day to receive two units. We got a driver from APLA to take Rog over in the morning, and the plan was that I would pick him up at two. Then I sat with fifteen dogs in a waiting room for an hour and a half, waiting to leave Puck off.

Arriving at UCLA, I expected Rog to be feeling spunky, since previous transfusions had always energized him so much. I knew Cope had spoken of checking out his cough and perhaps giving him a blood-gas test, but when I'd talked to Roger on the phone at noon he said the meeting with Cope had gone fine. I walked in at one-thirty and saw Roger sitting on the edge of the bed, all fresh-blooded, and I greeted him heartily. And hearing my voice, he looked over full of pain, and said in a tragic voice, "They're carving my tombstone."

One of Cope's colleagues had just been in to tell him his blood-gas number was in the sixties, which had always indicated PCP in Roger's case. They could only be sure with a bronchoscopy, but Roger categorically refused. He would not go through that test again, recalling the night when his throat had frozen and he begged us word by slow word, "Why is this happening to me?" If they felt so certain he'd broken through again, then they should just go ahead with treatment. I fell instantly into the support mode, promising him he wasn't dying: We *knew* this infection, we'd pulled him through three other bouts of it. I truly believed it as I said it, and didn't start obsessing about Roy Cohn and the breach of the AZT wall till later that night. They were admitting Roger right away, and I felt this helpless yearning to take him home.

Some of the agonies that burn in the heart forever begin as brief as snapshots. A nurse came to wheel Rog through the dozen corridors and bridges that connected the Bowyer Clinic to 10 East, and at one point we were on an elevator. Roger looked over and tried to see me five feet away, straining his one eye as if he were reaching for me, as if from a train pulling out of a station. That was the first time I ever suffered dying, and I can't even say what death it was. Roger's and mine both, to be sure, but something more as well. I understood then that the tragedy of parting was deeper than death—which only the very wisest have anything true to say about, like Mrs. Knecht across the street. "Here I am, Rog," I declared softly. He knew then that it couldn't be very far off, and I must've known as well but couldn't face it.

Yet that three-week hospitalization, the final extended stay, wasn't really so horrible. I was right when I said we knew this infection cold, and we stayed on top of it throughout, conquering four for four. In addition, we were blessed with a marvelous intern, Dr. Beal, who got who we were as soon as she met us. Her empathy and humaneness only threw into bold relief the gawky discomfort of the male interns we'd dealt with. Perhaps it was because Dr. Beal had gone to med school later than the rest and was ten years older than the kid doctors. She enjoyed us thoroughly, even after the incident. On the day following Roger's admission she was drawing blood, properly gloved of course, and she and I were chatting as she injected the blood into various culture mediums. Suddenly I saw her drop the needle; when she reached for it, it jabbed deep into her wrist. She stayed cool and went down for a gamma globulin shot, while I tortured myself with guilt that I was responsible, talking while she was working. She dismissed these thoughts firmly and did not go off the case and utterly minimized her own fears, though I could tell Cope was worried for her.

Gradually the hospital rhythms took the terror of death away, especially when Cope assured us Roger would beat the PCP. Except for that cry about his grave and the haunted piercing look in the elevator, Roger was fairly calm and comfortable, and for some reason didn't develop any drug reaction to Pentamidine. By the end of the first weekend his cough was already abating. Unhappily, I wasn't sleeping. I'd chosen this month to taper down on the drugs I took at night, and was weaning myself from the sleeping pill, figuring I needed the Xanax more, for anxiety. What I didn't know then, but which became clearer as the days passed, was that ribavirin had the side effect of insomnia. The night dose was like a jolt of caffeine, but I couldn't get anyone to corroborate that. Craig, who'd been on it for a year now, said he'd long ago given up sleeping deeper than an inch below the surface.

The tiredness made me punchy and weepy, especially when I would come back to the empty house. A few days after Roger went in, I arrived home and the phone was ringing. I stupidly picked it up, though by now I never answered without monitoring every call through the machine. It was Joel from Santa Fe, whom I hadn't spoken to since Leo died in April. The sound of his voice tied my stomach in knots, especially because the tone was so post-AIDS—by which I mean every word was full of his antibody-negative status and the putting of AIDS behind him, as he swore he would. About Leo dying he said, "Leo was ready to go. He didn't want the rest of us suffering anymore."

I suppose he must have asked about Roger, but what he seemed most eager to talk about was his new boyfriend, "who's very understanding about what I've been through, and helps me get on with my life." I was gasping with rage by this point, but all I said was: "I'm not going to be around long myself, and I don't want to talk to people without AIDS anymore." He hastened to say some drivel about not committing suicide for the sake of one's friends. "Have a good life," I told him, and hung up in the middle of his saying "I love you." In the middle of the verb, in fact.

By now Roger was registered with the Braille Institute, which had provided him with a special tape recorder and a catalogue of books on tape. He had listened at home to a specially prepared text for the suddenly blind, but in the main the catalogue was fairly middlebrow for our tastes. Yet Robbert Flick would go back and forth to the Institute, fielding tapes that Roger might like, and I recall the day when Roger tried one of Mary Renault's novels of the ancient world. When we were in Greece we'd both read books from her Athenian cycle and loved them. But when I came up that evening he'd abandoned it because he couldn't figure out how to fast-forward the tape. He'd gotten mired in the prefatory tables, listening to an endless chronology of kings and dynasties. I remember him making people laugh for days afterwards, telling how he'd been trapped in the lineage of Persia.

Yet the boredom factor was very real during that first week before his parents arrived, especially since he was feeling pretty well. He was strong enough to walk in the corridor, even to go outside in the plaza, though the latter entailed a tricky juggling of IV pole and wheelchair. But when I wasn't there he would usually be by himself, since most of our friends couldn't be with him on weekdays, even if the loyalists did drop by in the evening. I tended to arrive at the hospital myself midafternoon, but I can't remember exactly, because by then I was glazed with fatigue. Why did I let him stay there alone, with nothing better to do than sleep? I suppose I tried to carry on a couple of hours of business every day. It's a curious kind of guilt, wishing I'd known how little time was left. My seven or eight hours a day at UCLA seem so paltry to me now, and I must have wasted the rest of the day, since nothing comes back to mind. I'd gladly give a year to have any one of those days again, for I know precisely where I'd be, the whole twenty-four hours.

We decided—everyone decided—that I was so strung out and exhausted from lack of sleep that I must get away for a few days once Al and Bernice came to town. My parents had been pleading with me to visit them in Massachusetts, since they hadn't seen me in a year. I didn't want to go anywhere—couldn't leave Rog—though by now Sam was concerned enough about my hysterical fatigue to suggest my checking into a hotel, any hotel, just to collapse. In the happier part of the summer I'd mentioned to Rog that if things stayed stable we might get up to Big Sur for a long weekend, the way we used to. It wouldn't matter what he could see, because I would be there to tell him. Besides, we knew every trail and overlook in our sleep. Now, with friends warning that I was on the edge of a breakdown, I wistfully brought up Big Sur again. Roger pounced on the idea: clearly that's where I must go. I always liked it when he'd pull me off a fence and make a decision. I also suspect he colluded with Cope, who reassured me that all would be fine while I was away.

We decided I'd leave on Tuesday night, the day the parents were arriving from Chicago. Then they would have the house to themselves and be able to take care of Puck. During the weekend before, we talked about the trip and what I would see. Roger promised—I made him promise it over and over—that I could call him several times a day, and he'd send everyone out of the room and talk to me. I started to feel excited in spite of myself.

One afternoon, I walked in calling "Here I am," as usual. I realize now that I would announce myself this way as a counter to his blindness, but it's still the phrase I speak when I visit the grave, or sometimes when I walk into the empty house. As soon as he heard my greeting he smiled and declared, with a mixture of astonishment and tenderness, "But we're the same person. When did that happen?" As if he'd been waiting all day to say it. I agreed up and down right away, yet I've also brooded on it longer than almost anything he ever said. I think the reason for the "But" is that this was his answer to the darkness that told him he would die. But how could he die and leave me? How was it even physically possible to separate us now, with the two of us so interchangeably one?

I came home that night to find the goldfish dead in his bubbling tank. With the oddest dispassion, I gathered Schwartz, the tank and all the fish food, hauling them down to the trash barrel. And thought: If somebody has to die in this house, Schwartz, I'm glad it's you. It took me days before I could bring myself to break the news to Rog.

Fred was coming up to the hospital twice a week for their regular work schedule, and he and Roger coordinated the drawing up of the will with Esther. I was the problem here. Though Roger had been speaking off and on all summer about changing our wills, I couldn't figure how I wanted to set up the trust that would guard my work when I died. In the '80 will I'd appointed a fellow writer to be my literary executor, but he'd developed a certain contempt for my work, and I couldn't figure who else to trust. I let the weeks drag by, and finally Roger decided to go ahead with his own will. He would leave everything to me, but if I should not survive him by six months, then his half would go to his family. There were gay couples dying all over now, within weeks or months of each other, so contingencies had to be written in. Also, Roger didn't yet have the living will that Cope had mentioned eleven months before.

I recall how delicately Roger would speak to me about the will, always qualifying the gloomy portents. "But that's if you survive me," he'd say, explaining some detail. "If I survive you . . ." He knew I couldn't handle the death part. He worried to Dr. Martin during a therapy session at the hospital: "Paul says he can't survive without me." An accurate quotation, I'm afraid, and I'm grateful to Martin for allaying his concern: "Of course he can. He'll do fine." Even if it isn't so, I'm glad he said it.

Roger had been in ten days when Al and Bernice landed on the evening of Tuesday the sixteenth. I extracted a final promise from Rog that he would stay strong and stable. Then, with enormous trepidation, I got in the car and headed north, reaching Santa Maria at 1 A.M. We always left for the seven-hour trip to Big Sur late at night. Roger would sleep while I drove, and we'd stop in Santa Maria for a few hours' sleep and have breakfast next morning at Morro Bay. Wednesday I started up the coast road in cloudy weather, three hours of the staggers of Route 1. I've never felt quite so schizoid as I did coming into the soaring calm and noble immensity of Big Sur, missing my friend and feeling as if I didn't deserve this beauty anymore. I stopped at a turnout south of Partington Ridge and realized as I made my way along a trail to a waterfall that Roger and I had never walked together in exactly this place, and I fell apart. I knew then the only way I would get through the three days was to realize I had come here to say good-bye from both of us.

I stayed at the Big Sur Lodge, hiking every morning to the mouth of the Big Sur River, where it spills out onto Andrew Molera Beach. Molera is where I took the pictures of the two of us in October '83, the weekend after Cesar's diagnosis. The wild beach sweeps five miles in a curve like the quarter-moon, banked by headlands bare as Scotland. If you walk a half mile along Molera, there is nothing after a while but where you are, and I'd hole up under the bluffs and sun naked. One morning I wrote with a finger in a drift of powder sand in a hollow below the bluff, "P & R," just so I could tell Roger when I called him from the phone booth at the lodge. I left our mark, I told him.

In the afternoon I'd go down Sycamore Canyon to Pfeiffer Beach, where the ocean roars through tunnels in the rocks. In between I walked in the redwood groves. But three times a day I'd home back to the outdoor phone booth and call the tenth floor at UCLA. I'd spill out all my travels and tell Roger about the strange double nature of it all, how I would be exalted one minute and crying the next with fear. Then I'd ask all the rote questions about his numbers and the doctors, and he'd make me easy and send me out for another hike, loving every description, for I was the last of our eyes. He talked me through the whole trip, and once when I couldn't bear the pain of being far from him he said, "I miss you the same way, darling. But there's part of me that's rooting for you to have a good time. So try."

I tried. Sometimes I'd call the hospital and he'd be asleep. Perhaps his mother would talk to me, or a nurse would tell me to call later. Then I would walk in circles till I could call again, unsatisfied till I had the reassurance direct from his lips. But I did sleep eight or nine hours a night, with naps in the afternoon as well, and the best call was always at 9 P.M., before the switchboard shut down at UCLA. While we talked I'd look up through the redwoods at the billion stars, my breath smoking in the autumn cold. He'd laugh and tell me jokes from the old Jack Benny tapes his cousin Ruth had sent him. The last night I told him I was terrified. "Of what?" The future, I said. What's going to happen to us. And he replied in the mildest voice: "You just come back, okay? And then we'll continue our ongoing togetherness."

Saturday morning I left the lodge and drove north to Carmel to pick up the freeway, and for the whole twenty-five miles I took my last leave of Big Sur. It had been ours for a decade, and I didn't want any more of it. I took the inland route down the spine of the state, the golden hills so arid they gleamed like platinum, and stopping every two hours to check in. I hit brutal traffic coming into L.A. and arrived fried at the hospital at six-thirty. When I walked in the room his parents greeted me with a cheer. Roger, lying half asleep in bed, was so pleased and excited that all he could do for a moment was moan with pleasure, rocking back and forth in a motion akin to wagging.

We had finished the *Apology* by now and moved on to the *Crito*, the dialogue named for the friend of Socrates who visits him in prison and lays out a plot by which the philosopher can escape into exile. It's a very problematic piece to read, listening as Socrates decides he cannot flee the state that has put him to death without destroying *all* the laws of Athens in the process. Meanwhile, sublime to ridiculous, I'd been working for some weeks writing the novelization for *Predator*, the upcoming Schwarzenegger opus.

While I was in Big Sur I decided I simply couldn't make the November 1 deadline, and I curled up with Roger in his bed in room 1010 and said the project was too stressful. Though he'd encouraged me to keep working throughout his illness, now he said: "Paul, if you want to pull out of it, go ahead. I want you here with me now. Who cares about all that?" With a lifetime of Puritan ethic behind me, I'd never pulled out of any project, no matter how wrongheaded. For the first time I actually considered it, and we discussed it again and again. When I decided to go ahead with Arnold and the alien, I had to promise myself I wouldn't let it make me crazy. I asked for more time from the sweet-tempered editor at Berkley, to whom I never mentioned Roger's illness till after he died. Now it seems like yet another portent, his wanting all the time he could have with me, he who was so unpossessive.

Before his parents left, Al once again paid him the highest compliment about his relationship with me. "You boys are the best friends I've ever seen," he said. "You're like Damon and Pythias." It's a long way for a man to come who couldn't look me in the face for a year after Roger finally told him he was gay. A century after Socrates, Pythias was condemned to death in Syracuse but wished to go home and settle his affairs. Damon, his friend, took his place in jail, agreeing to be executed if Pythias didn't return. Of course he returned, and the tyrant who'd condemned him was so moved by the friendship that he released them both. This is not a myth.

Al and Bernice promised they would be back November 22 for Roger's forty-fifth birthday, which coincided with the fifteenth-anniversary dinner of the Gay and Lesbian Community Services Center. One of the themes of the Center dinner this year was the uniting of families, and people were being encouraged to bring along their parents and siblings. I remember Al and Bernice leaving us cheerfully in room 1010, with not a whisper of good-bye.

One night I read an essay to Rog about Francis Parkman, on the occasion of the republication by the Library of America of his history of the French and Indian War. Parkman was virtually blind and in great pain during the long ordeal of writing his book, and he had to work with a kind of iron grid over his paper to guide his pen. We both choked up at his fortitude in the face of daunting illness and disability. But if there was a certain weepy sentimentality to us now and then, Roger could be sharp and mordant as well. He had a call from a friend back east who didn't know what to say, and when he asked, "How are you?" Roger replied, "How am I? Read the Book of Job."

He began to run a fever during the third week in the hospital, and Cope and Dr. Ahn decided that the catheter had become infected. A typical problem, apparently, though it was the first we'd ever heard of it. Roger would have to go into surgery Friday morning to replace it, but the good news was that he could come home Saturday. I remember taking him down Thursday night for an x-ray, always easier than waiting for an escort. The man in the wheelchair ahead of us was shrunken and covered with lesions, all alone, and for once I was glad Roger was sightless. I don't remember him being anxious about the surgery, yet when Richard Ide came up to visit him Friday morning and arrived an hour late, Roger rebuked him in a wounded tone: "Richard, don't do that to me again." Richard is never late for anything anymore. I accompanied Roger down to the operating room, just as I had at Jules Stein in April, but this time it wasn't so overwhelming. I squeezed his hand at the door to surgery and said, "You be okay now."

I wanted to be with him that evening, in case he was feeling "wimbly" after the operation, but I'd promised Rand Schrader that I would host an important meeting at the house that night. In order to develop a text for the video presentation at the Center dinner, I'd agreed to gather some key people who'd been there since the beginning to give us the narrative line. No history had yet been written down, of the

Center or any other aspect of the gay movement in L.A. I was torn about leaving Rog, but he swore he'd be fine and felt the occasion was too important to miss. Besides, he was so looking forward to coming home next day that he appeared to shrug off any aftereffects of the surgery. Yet I was the same as he was now, jealous of any time that was stolen from us.

About six people gathered in the house on Kings Road, including psychologist Don Kilhefner, the shaman granddaddy of the movement in L.A., and Steve Schulte, director of the Center during its turmoil years of the late seventies and now mayor of West Hollywood. Kilhefner told the story of the "gay survival committee" in the years following Stonewall, which led in 1971 to the opening of Edgemont Liberation House, the first truly public gay environment in the city, gay people helping each other. A clinic opened in 1973, and the Center gathered clout and respectability as it made inroads in the gay middle class, most of whose members were still officially in the closet.

Everyone gay starts out in the closet, of course. The Center, by proclaiming "Gay" on the building on Highland, had made a stand about coming out, though what Kilhefner called the "inner coming out" would take more time. There was a difference between what we were and who we were as gay people, among other things that we were a people and not just a movement. It was extraordinarily moving to hear them talk about how far we'd come, despite the calamity, and about what we had marshaled to fight the calamity with.

Cope ordered a transfusion for early Saturday morning. When I called at ten, Roger reported with a flush of pleasure that he'd declared to Cope when the doctor came to check him out, "Dennis, I'm feeling optimistic." Cope replied, "I think there's reason to, Roger." So my friend was beaming when I came by at noon to pick him up, and he even laughed about a student nurse who'd treated him like a creature from outer space. We drove home in cool September weather, and the ritual of homecoming was an aching delight, from the dog turning inside out and whimpering to the tramp upstairs past the coral tree. I remember when we got into the back bedroom and sat down together on the bed, facing the garden, we laughed to think he was home, our heads tilted against each other as we savored having come again through fire. It was perhaps the last moment of full joy, but I can still taste the triumph of it.

Admittedly the boost from the transfusion didn't linger as it had in the past. On the other hand, we were accustomed now to the uphill fight for strength after PCP. We were fortunate to be able to have Scott reassigned to us. Though he had to move on to other cases when Roger went in, he'd been mostly working as a substitute and was eager to return to us. Our priority was the restoration of the quiet sanctuary of the summer. Right away we were back to mornings in the garden, Roger sitting in the dappled shade while I read him the paper, afternoons with Fred, suppers prepared by Scott, and Plato in the evening. I finally unpacked the boxes piled in the living room since February, when the office move had taken place. I even spent a couple of evenings going through drawers of snapshots, describing them to Roger as we called up a raft of memories.

The restoration was real and tangible, and went on with gathering confidence for the next week or so. I was even bitten with the desire to bring the house back up to form. The specific goad was the sofa in the living room, whose re-covering we had abandoned when Roger first took sick. Thus the upholstery bore a year and a half more of Puck's grime, and was smelling exceedingly doggy. Not wanting Rog to lie down on it anymore for fear of germs, I announced that we would re-cover at last. I decided we would also acquire a couple of easy chairs, and under the new rules Puck would be banished to the floor, where he belonged. Roger encouraged me in this sudden intense enterprise, as I called for estimates and hauled home fabric samples so he could feel them and approve them. I know I felt an extreme urgency to make the place comfortable for him now, no time to waste. I told every workman ASAP, or STAT, as they say in the medical labs.

Because I'd planted late in the spring, the gardenia was still flowering two or three blooms a week, a rare thing for October. Gardenias were usually finished by August at our place. I took to bringing a blossom in and placing it in a dish of water on Roger's nightstand, crowded by pill bottles and IV material though it was. I was actually returning a favor here, for Roger had been for years in the habit of bringing the occasional gardenia in and leaving it in a dish on my desk. Before setting it down I would give him a whiff, and he'd purr, "I love you so much." But that was something he said more and more often now, as I would dart into the room with some fattening drink or chaotic question about our finances. You cannot ever say it enough, of course, but he spoke it now with a rare savor. More often than not I'd parrot it right back to him, but I had my own schedule of telling how much I loved him. It wasn't a contest, nor was it ever perfunctory. For both of us it was simply a statement of fact, as much as to say *I'm alive.*

This is not to say there were no moments of pain and loss, nor too much time to brood all alone in the darkness, in spite of all the visits, the Mozart tapes and the massages. I think he would comb the rich hours and countries of his life, till the tragedy would suddenly break over it all like a tidal wave, then he would cry out. The most painful of these, the one that cut deepest into me, was a moment in early October as I came into his room late one afternoon. He was looking out toward the garden, though by then he could scarcely see the light. He cried out softly, in an agonized voice, "What happened to our happy life?"

I think about almost nothing else now. But at the time, I said what I really believed: "It's here, Rog, it's right here. Because *we* are." I required nothing else, but then I was not hobbled and assaulted as he was. I did not have to inhabit the dark and remember the voyaging, the comradeship, life engaged to the full. In addition to which he had no manic phase as I did, to fill up with all the IV tasks, or the record I leave behind here. The memories that broke his heart with being gone are the ones I live with now, of a life so happy it hurt sometimes, like the meadow Miriam couldn't take home in her arms in *Sons and Lovers*.

It all goes one way now. Difficult in the extreme to steer the course of the last weeks without being thrown off by a feeling that I somehow failed to keep him alive. I punish myself for lack of vigilance, thinking if I'd got him to the hospital sooner I would've pulled him through the final infection. Pulled him through to what? one could reasonably retort. They tell me I wouldn't have wanted the lingering weeks of devastation, the final explosion of ravages that drags people in and out of comas, pleading to let it end. And I've come to understand, intellectually at least, that our triumph was in what we *did* do, keeping him alive and alert and on our island till the end. Yet I can't think of almost any moment of October without feeling helpless, like flinching in the glare of the final air burst. But how was I to know? Then I knew nothing about death, and now I know everything short of my own.

It must have been midweek the second week of October that Scott greeted me when I woke up with the news that Roger had been incontinent in the night. Incontinent is so curiously Protestant a term that it puts the reality at some distance, and I recall not even asking him or Roger what sort or what degree. Roger used the urinal bottle in bed much of the time now, at least once an hour, since we were filling him with so many liquids. I think it was misplaced modesty on my part, or my sense of how modest Roger was himself, that kept me from probing. So he'd pissed in bed, so what? After I went to sleep at 4 A.M. there wasn't anyone to monitor him till the nurse arrived at seven-thirty for the morning IV. Accidents happen.

Jaimee would call at eight-thirty or nine at night, and while he was talking with her I'd often slip down to the gym for twenty minutes on the Lifecycle, being as the place closed at nine-thirty. A couple of evenings Roger was unhappy about my going, disappointed even, because usually by the time I returned he'd be ready for bed. He was so weary since the hospital that I'd sometimes have to bring him his toothbrush bedside, and it was all I could do to coax a weigh-in out of him—145 and holding. This one night he was on the phone in the study, and I swore I'd be back in a half hour, before he knew it. When I returned at nine-forty-five he told me he'd gotten lost in the house, wandering around disoriented till he realized he was in the front bedroom and felt his way back to his own. I put it all down to the disorientation of blindness, purely directional, and didn't connect it up with his getting lost by the pool equipment. He was on Haldol now, which did after all ease such problems, and perhaps even mask their depth.

I don't know in any case how much was a manifestation of dementia caused by viral invasion of the brain, since there wasn't anything dramatic about it. He was fully coherent, if spending a fair amount of the day sleeping, but this was also the time when he was getting the greatest pleasure listening to various tapes friends had brought him. Sometime during those last weeks he listened to *Julius Caesar* straight through. He hadn't read the play since tenth grade and was thrilled by its eloquent tightening web, talking it over enthusiastically with Richard Ide, our resident Shakespearean. He chortled through *The Importance of Being Earnest*, which Marjorie brought him, and was spellbound by a tape the photographer Holly Wright had made of the "Overture" to *Swann's Way* in French. It was impossible for me to focus on the diminishing of his mind, with him talking about literature with such evident delight and lucidity.

And yet I was manic and busy and didn't pay close attention. Saturday the eleventh I went out and took care of a groaning board of errands. I happened to pass a new postmodern minimall with a restaurant opening called Beau Thai. I could hardly wait to get home to tell Rog, who loved the idiot puns of California signage. But when I related it to him as I served him lunch, he made me repeat it and didn't get much of a laugh over it. Win some lose some, I thought. Then when Richard dropped by an hour later, Roger asked me to tell him the Beau Thai pun. And when I did, Roger commented, "Can you believe I didn't get that?" It was the one moment when he seemed to have a conscious sense of something not being quite right, but none of us really picked up on it or knew what to do with it. How do you factor in the missing of a single pun? That night we finished the *Crito* and launched into the *Phaedo*, Plato's dialogue about the spiritual life and the immortality of the soul.

Sunday afternoon the twelfth, Susan and Robbert brought over a birthday cake for me, since they would be swamped at school on Thursday, my actual forty-first. I recall we sat out on the front terrace in clear October light, and Robbert and Susan asked Roger a list of questions about a loft space they were thinking of buying. They'd never owned property before and wanted advice about how to talk to a banker. Roger was utterly in his element as he laid out a set of options for them, promising to make contact with our banker in West Hollywood. He was logical and patient, tuned to them alone, totally absorbed.

That night he said to me while we were puttering in his room, "You know, you're the most beautiful man." I grinned: "You mean physically?" He nodded with mock gravity: "Oh, yes, I always thought so." I've never seen anyone do that, spending all his endearments and giving voice to the best he knew in people. Over and over I'd hear from friends who talked to him in the last weeks—how they would call to try to stumble something out, and Roger would turn it around and want to help them with *their* problems. Each time he lobbed one of his encomiums to me I was so stirred and touched I didn't stop to think it was any sort of final testament. On the contrary, at such moments he could hardly have been more alive.

In the early years he was never as impulsive as I about saying he loved me, whereas I would spend those coins like a profligate, having waited all my life to speak such things out loud. Once in the summer of '80 I was in New Orleans researching a script, and he tried to tell me on the phone one night how empty he felt without me. I wasn't very patient, apparently, replying, "I'm probably not the one to talk to about it, Rog, because I just miss you." When I read his '80 journal after he died, I found a note to me at the bottom of that day's entry:

> *Paul—the important thing to say is this: with you it's been the best—the best years and the most love.*

The plain certainty of his words in the last two weeks was like that, their affirmation whole, with the directness of a heart burned clean of extraneous feeling.

He was suffering consistently now from fevers and drenching sweats, and I would have to change his pajama top at midnight and at two, sometimes again at three-thirty, before I went to bed. He had also been off AZT for over a week, but his white count hadn't bottomed out very low and was inching up again. He'd be back on it within a few days. I think there was another announcement from Scott about his wetting the bed one morning, but I definitely remember Tuesday at noon, as I was getting ready to go to Alfred's. Scott was serving him soup and a salami sandwich in the dining room. For months he had favored deli for lunch, which I applauded for being high caloric and proof of a hearty appetite. But now he vomited when he sat down at the table, and then as he moved blindly to go through the kitchen into the bathroom, he had a squall of diarrhea in his pants. I was rattled, and Scott said he'd take care of it, so I left for my meeting. As soon as I got home I put in a call to Cope, who would usually get back to us within half a day. He didn't this time, though Roger seemed all right and had no complaints outside of the fevers and sweats. And they were a constant that didn't tell us much anymore.

Yet I know I was frightened and bouncing off the walls that night, because at one point, as I ran around preparing the night dose of IV,

I started crying and hyperventilating with anxiety. What is it? he wanted to know. "I'm so afraid," I told him. "I want you to be okay. I don't know what to do." And he spoke an ocean of reassurance, curled on his side in bed: "Don't worry, darling, I'm still here." Twice more in the difficult days that followed he'd hear me blubbering over one misery or another, and he echoed again, "I'm still here." Letting me tuck in beside him and unleash a flood of grief.

Wednesday we had our hour in the garden in the morning, and when he sat down for lunch he was sweating buckets. I came in to say I'd be back from Alfred's at three, and he tilted his head as if he wanted propping up, with a small moan of weariness. I hugged him and told him everything was fine. Later that afternoon Dr. Martin called to find out if he should come by for the Wednesday appointment. Rog was in bed and shook his head no, he wasn't up to it. He hadn't been up to it the previous Wednesday either, as I remember, but he chatted with Martin for a few minutes. Then Martin asked to speak to me. He said, "I just asked Roger if he was sleeping a lot, and he answered, 'Yes, I'm sleeping for everybody.' " Martin let the statement make its own conclusion, that Roger was drifting off mentally. But I had the prior experience with Martin about the "brain involvement" and wouldn't give it the weight he did. Scott and I were feeling that Roger's symptoms were just like the dehydration episode in June.

And the day would sweep us on because there was so much still to persevere in. Right after the Martin call was the IV dose, and all through that hour the two of us talked shorthand. Or we would hardly talk at all and just be together, unchallenged by any outside view, psychiatric or otherwise. If it wasn't exactly twenty-four-hour care we were giving him now, it was at least a twenty-four-hour situation. Yet I didn't mind the constant part, for the seamlessness of the day—doses and meals, the night walk, the weigh-in—became its own kind of walled city, magically protected.

Thursday was my birthday, and I reluctantly agreed to celebrate it for everybody else's sake. Alfred took me out for lunch, then we headed to Culver City for a meeting. By the time I got home it was four-thirty, late for the afternoon dose. Joe and Vince, two dear friends who had just celebrated their fiftieth anniversary, were visiting at the time, and Roger seemed all right. The next hours were back-to-back with phone calls from our two families, the rote greetings, gifts to thank them for. Richard Ide came over for cake in the evening, and as he had the previous Christmas, he brought on Roger's orders several presents. Though feverish and wilted, Rog was clearly very pleased that we'd pulled together a bit of a party.

And yet Richard says he knew that night that Roger was fading quickly. Winding down, as someone put it to me the other day about a friend who's entering the final spiral chamber. I can see it immediately in others now, but I couldn't in Rog. Richard says he finally noticed the scope of my denial. I remember that Jaimee called, and Roger begged off talking to her because he was visiting with Richard. In fact he was trying to tell Richard a story about his niece Lisa, and kept having to call to me in the next room to untangle the details. Richard knew then. I just untangled the details and went on talking to Jaimee.

Cope called after that, and we decided in the absence of any specific symptoms to sit tight till we saw the results of the next day's blood tests. I told Jaimee and Cope the same thing, that I planned to spend all my time with him over the weekend and graph everything that happened, making sure he was properly hydrated. We all had credence still in the urgency of liquids, since dehydration involved its own disorientation. When I hugged Roger in bed that night I said, "You're my birthday present, okay? You're all I need." And he laughed and murmured, "My best best friend."

Somehow we got over to UCLA on Friday for blood tests and an eye appointment. I insisted on a wheelchair because of all the walking, and I remember we had to see one of Kreiger's colleagues, since the doctor was out of town. The substitute was a man who treated us brusquely, as if this case were too hopeless for his time. Nothing new to report: Though the infection was still stable, Roger couldn't see the chart at all and was too feverish to care. We made an appointment for Tuesday, when Kreiger would be back and we could schedule laser surgery for the cataract.

For the rest of the day and the evening I spent all the time with Rog, monitoring him and constantly asking how he was. Fine, he'd say. I wonder now if our very closeness kept me from seeing a pattern of dislocation and withdrawal. After all, it was so easy to leave things half said, having finished each other's sentences for years. The care system we had evolved was so entrenched in teamwork, every milk shake. I think that at the back of my mind I couldn't conceive of him in mortal danger because his weight was stable and he was eating well. Friday night, I remember, we talked about the discovery, just announced by *National Geographic*, of Columbus's landfall in the Caribbean. But I wonder now if I didn't carry the burden of the conversation for both of us, just as I must have done all week with the Plato.

Friday night I was seized with the old guilt. "I feel so terrible, Rog," I said as we got him ready for bed, "that I made such a mess over Joel." He seemed surprised that it still troubled me, and he reached for my hand. "Paul, we got through all that. It's not important anymore."

When I went to bed at three my panic had subsided, for we'd held to a neutral day and would have the whole weekend to rehydrate him and bring him back to strength. Weekend mornings he'd stay in bed later because Scott didn't come, and the nurse would let herself in and be out by nine. I went in when I heard her leave and asked him if he wanted to come and cuddle with me in bed for a while. I led him to the front bedroom, half asleep myself and telling him everything was going to be fine. We dozed for a half hour or so, till he abruptly sat up, and I woke and asked if he needed the urinal. I think he said yes, because I went and fetched it from the other bedroom. He stood up and held the bottle in his hand, poising his dick at the lip of it. But nothing came out.

"What's wrong?" I asked. "Go ahead." He seemed to be concentrating fiercely but also to be puzzled, and then suddenly he did the strangest thing. He cocked his head, bewildered and curiously intent, like a deer not quite sure if he's just heard danger. Then quickly he whipped the urinal bottle around behind him and started to defecate into it. "What are you doing?" I asked him frantically, shaken by the dislocation and irritated at being awakened. I took the bottle and tried to lead him down the hall to the bathroom, and he stumbled and fell to the floor, staring up at me, staring blind. "Get up!" I cried, trying to drag him to his feet. He looked stunned and oddly shell-shocked, and troubled at how upset I was.

I was gnashing my teeth with fear, but after a moment I forced myself to be calm and began to talk gently. I helped him to his feet, cleaned him up and got him comfortable in bed. By now he was saying he was all right, he just wanted to sleep. So did I. All I wanted to do was sleep, and that's what I did for an hour. In that moment when I should have had us in the car on the way to the hospital, I couldn't cope anymore.

I think I was in shell shock myself from then on, but you don't somehow notice the gaping hole in your own head. We had breakfast, and Roger seemed weary but otherwise himself. I'd placed a call to Dennis Cope as soon as I got up, but as it happened he had his own crisis that morning—broke his foot and had to come into the emergency room to have it put in a cast. The doctor who was on call for him didn't get back to us till midafternoon, and by then Roger and I had had a walk up Harold Way, and he took a call from Tony Smith in Boston. I greedily drank in the reassurance of hearing him laugh and talk with Tony. When the on-call doctor finally checked in, he listened to what I had to say but didn't feel I should rush Roger over to the hospital. As long as he seemed all right now, Cope would be in touch with us tomorrow, and surely Monday was soon enough for an examination.

Then he read off the numbers from the previous day's blood tests, and the white blood count was 800. But that couldn't be. It had bottomed out just under 2,000 and was on its way up. Roger hadn't started back on AZT yet, so how could it have come down? The doctor had no answer; Roger wasn't his patient, and they tend to dump such disparities on the "weekend problem," when the labs are notoriously inaccurate and presumably staffed by chimpanzees.

After the four o'clock dose Roger wanted to nap and told me to go to the gym. When I got back an hour later I found him feeling his way in the dining room in the dark, drenched with sweat, as if he didn't know where he was going. We changed his clothes and had dinner. The incontinence wasn't consistent at all, for he used the urinal bottle or the toilet the rest of Saturday. I was scared about the white blood count, however, and automatically put on a blue mask to protect him.

Then the evening ended with a bold stroke of reassurance. After all those sweats I'd given him cursory sponge baths, but now I asked if he wanted to wash his hair. "Oh, what a good idea," he said happily, and we traipsed into the bathroom. I swathed him in towels to protect the catheter, and he knelt and bent over the tub while I shampooed him. There was something unimaginably secure in the ordinary rhythm. He was so refreshed afterwards, laughing as I bundled him dry, more animated than he'd been in days. When I had him back in bed in clean pajamas, we talked about the queer incident of the morning, shitting in the bottle. I apologized for yelling at him, and he said he hadn't been sure what to say in the wake of my panicked reaction. It's hard to explain how we normalized and explained it all away, but the serenity and closeness of the night made all the earlier turmoil seem like a bad dream.

At 10 P.M. Alfred dropped over for an hour. More than ever, I hated leaving Roger, with the two of us connecting so effortlessly. But Alfred and I had a script deadline, and I'd promised him a couple of hours over the weekend. "Go ahead," said Roger, "I'll be fine." Later, during the midnight dose, everything was peaceful as I gave him pills and made him drink lemonade. Then I watched over him till four, working on *Predator* while I sat by his bed.

Sunday was bad from beginning to end, but all in a minor key. There was nothing so obviously off kilter as the incident with the urinal bottle, and I never had the experience of hearing him speak in non sequiturs. Most of the day was no more harrowing than being confined to the house, though he was fevered and sleepy and didn't talk much, mostly answering yes and no. But we went through the rigors of being up and about, as if I could keep things stable by fanatic adherence to our hard-won schedules. Trying to hold back the flood, making the beds as the missiles arrive. When Cope called midafternoon, we discussed the situation from various angles. Still he thought it would be all right to wait for the regular Monday noon appointment, especially since Roger wasn't running a very high fever that wouldn't break, like the one that sent us to the emergency room in June. I said I thought I was keeping Roger stable, and I don't remember what we said about the white count.

We sat out on the terrace, going through mail. The flier had arrived from Florida Orange Growers, from which Roger would always order a bushel of grapefruit for my parents for Christmas. I remember us talking about that and me filling out the order, "Love, Roger." Rand Schrader dropped by to visit, since I'd told him the previous night that Rog wasn't up to the regular Sunday breakfast. Rand knew right away that things had taken an ominous turn. Roger was barely there, he says, confused and no longer able to have a real conversation. "It was like we were talking about him in front of him, instead of him engaging with us."

But when you are backed in a corner, you look for any sign of light at all. He was able to come to the table for supper, however wilted. I know we had chicken, because I remember it was my proof of how functional he was, that he could eat it. I also remember the leftovers in the fridge after he died, mocking me with all my rituals of feeding. I spoon-fed him a bowl of applesauce for dessert. Then we had a last walk up Harold Way, where he leaned on my arm but still managed a hundred yards uphill, with the misty view out over the city lights that only I could see.

When he wet the bed later I changed the sheets, maddened now, as if I thought he was doing it deliberately. As I put down plastic under the sheets he asked what I was doing, then shrugged and looked sheepish. Jaimee called, and I told her it had been a difficult day. I was worried about the white blood count and what would happen tomorrow, ". . . but Jaimee, he's all right. I don't think it's anything dangerous."

And Roger perked up beside me in bed and shook his head: "No, of course it isn't." So it wasn't—not yet. However horrible what would come in the next two days would be, he was feeling peaceful enough just then, taken care of and going along, no pain or fear. That turned out to be all we could give him, but we brought him that far. I gave him the night IV and stayed up again until four, watching him sleep peacefully, waking him every hour or so to give him water and change his sweat-soaked shirt. Perhaps because I have always worked by midnight oil, I never believe that anything truly bad can happen on the night watch.

Scott strode into my room at nine and woke me. "You have to get Roger to the hospital," he said. I was blurred with sleep and tried to explain we had an appointment at noon. "He's rigid," Scott said. "His temp's 103, and he can't talk to me." So I ran in, pounding with adrenaline now, and talked softly to Rog, coaxing him to respond. He seemed to relax and murmured "All right" when I said we were off to the hospital. Scott was furious because the morning nurse had obviously ignored the problem, leaving Roger to lie in fouled pajamas. I remember that as Scott brought him downstairs to the car, Roger was walking on his own but seemed wooden and stiff.

Then as we drove over to UCLA, me patting and reassuring him, the fever began to break, and he started to sweat. At a light in Beverly Hills I took his temp with a throwaway thermometer. It was down to 100. He seemed more focused, and we began to talk. As we arrived in the emergency room parking lot I told him I'd go get a wheelchair, and he said, "That's not necessary." But I got one anyway, put a mask on him, and wheeled him in to the registration desk. The nurse droned through a list of questions, asking what was wrong, and I said, "He's had a high fever, and he's a little incoherent."

"I am not," piped up Rog with a certain stubborn pride. I smiled at his tenacity. It was the last full sentence I ever heard him speak.

We waited in an examining room for Cope. Roger's temp continued to hover toward normal, and I gave him an Ensure to sip through a straw, certain that whatever crisis had erupted this morning had passed. But then, when Cope came in, in a wheelchair of his own, Roger couldn't answer most of his questions. When's your birthday, Roger? "November twenty-second," he replied, and my heart leaped with triumph. But then he was silent when Cope asked, "Who am I, Roger?" I could see Rog straining to answer, trying to focus, and when the next question came, the intercom in the hall announced some neutral business, paging a doctor, perhaps. Roger, struggling to speak, began to parrot what the intercom said.

Out in the hall, Cope told me they would run the usual battery of tests—bone marrow, spinal tap, x-ray. He suspected it was either cryptococcal meningitis or the AIDS virus in the brain, and the one was treatable and the other not. I said I would stay with Roger till the tests began, then I would go back to the house and pack him a bag and meet him on the tenth floor when they admitted him in the late afternoon. I stayed by him through the x-ray procedure, talking and holding his hand in the glaring hallways. When they brought him back to the ER for the spinal tap, I told him I'd be back in a few hours. He said "Okay," or perhaps just murmured yes; I don't think he could talk whole words by then. But I didn't realize, and I still don't know why I left.

I went home and stared at his bed, where the sheets still swirled with the shape of his sleep. Then I stripped and jumped in the pool, though it was freezing cold in late October. I made calls to various sources in the underground to check out the central nervous system, pulling together anecdotes about meningitis. I took a nightmare nap, packed an automatic bag and got up there about five, when they were just putting him to bed in room 1010.

He was clearly miserably sick when I walked in, but I'm sure he could still answer me yes and no, because I had no sense of his speech center having been affected. I got him to drink another Ensure and ordered him dinner. Then two interns came in and stood by the side of the bed and announced he had cryptococcal meningitis. They would begin treatment right away with amphotericin B. Rog didn't answer them—didn't move—and I dismissed them and told him how lucky we were it was treatable. As if I had no other choice but driving forward into the teeth of the gale. I remember feeding him soup, and him looking up at my face with a kind of stillness in his own, yet full of an indescribable yearning. I don't know how dim his sight was then, or how locked the muscles of his face, but I felt him looking at me with a heartbreaking immediacy. He dutifully ate all the soup, then quietly vomited it up.

Cope came in in his wheelchair and went over the diagnosis again. Now Roger's eyes were wide with terror, and he could barely respond with a murmur. Cope promised him we'd bring him through it and was at once forceful and infinitely kind, after his fashion. Then he spoke to me quietly at the foot of the bed, saying, "Paul, I think this is worth fighting because the quality of life he has at home is worth it. But it wouldn't be the worst thing if this were the one that took him."

I stared at him, unable to hear it, and made him declare again that the medication would restore him. Then, as we waited for the pharmacy to send it up, I made phone calls to the families, especially to Al and Bernice, informing them that Dr. Cope thought they should fly out. I didn't know they'd asked him to alert them months before if the end was near. Though Sheldon tried to put them off till later in the week, they called me back to say they'd be on a plane in the morning.

Amphotericin B is administered with Benadryl in order to avoid convulsions, the most serious possible side effect. It was about nine or ten when they started the drug in his veins, and I sat by the bed as nurses streamed in and out. A half hour into the slow drip, the nurse monitoring the IV walked out, saying she'd be right back, and a couple of minutes later Roger began to shake. I gripped him by the shoulders as he was jolted by what felt like waves of electric shock, staring at me horror-struck. Though Cope would tell me later, trying to ease the torture of my memory, that "mentation" is all blurred during convulsions, I saw that Roger knew the horror.

I kept waiting for the nurse to come back. Why didn't I press the call button? Did I think the horror was *supposed* to happen, another thing to endure? We had stood in the fire so long that burning was second nature now. I sat there frozen, holding him for endless minutes, trying not to cry as I told him I was with him. *Here I am, Rog*— but with all the cheer and exultation drained away. When the nurse returned she looked at him in dismay: "How long has *this* been going on?" Then she ordered an emergency shot of morphine to counteract the horror. When at last he fell into a deep sleep they all told me to go home, saying they would try another dose of the ampho in a few hours. I was so ragged I could barely walk. So I left him there with no way of knowing how near it was, or maybe not brave enough to know.

I went home and called Jim Corty, Sam and Craig, who all reassured me the ampho would kick in. They were full of cases that had shaken the stranglehold on the nervous system, and the convulsions were to be expected. He'd come back; they swore it. I sat at the desk unable to sleep, working numbly on *Predator* for an hour or so. I called UCLA at two and again at three. They said they were having trouble keeping his fever under control, but otherwise he was stable. You force yourself not to think about the pain, where it hurt this time and how bad. I cursed myself for not having a private nurse with him and ordered one for the morning. But that was all: I went to bed certain he'd be responding to the drug within a few hours. I would not see the dying.

I started calling at seven-thirty, as I always did when he was in the hospital, checking in every half hour till I got up. The private duty nurse said he was resting comfortably, and I understood he'd had the first full dose of the ampho early in the morning. It turned out she was merely stonewalling me—didn't know who I was, certainly didn't understand that I was the one who had hired her. In fact they'd tried the drug at five, and he'd had another convulsion, so they stopped it. Nobody ordered it started again. It waited for someone to make a stink, and I wasn't there to make it. The unraveling was on every side.

I got up and was getting ready to head over, but taking my time, in no rush at all, when a call came in from Michael, Roger's brother-in-law in Chicago. He'd just talked with a neurologist, who told him that if they administered the ampho even a few hours too late the brain might have swollen so much as to cause permanent damage. I didn't really hear the awful details, all I heard was the primal urgency in Michael's voice. Within two minutes I was in the car, roaring up Sunset Boulevard.

There's a sucker speed trap just before the Beverly Hills Hotel, and I was going fifty in a thirty-five zone, but the problem was the Datsun. They don't stop Jaguars in B.H., not wanting to ruffle the core constituency. The cop had me cold with his radar gun and pulled me over. I erupted out of the car, screaming that my friend was in a coma—only time I ever used the word—pleading to be taken to the hospital. He said nothing, ignored me completely, and methodically, very very slowly, wrote out the ticket. I went bullshit, shrieking at him at the top of my lungs. Did he enjoy making people suffer? I'm certain he would've been within his rights to arrest me on the spot, but he didn't even tell me to shut up. Coldly he handed over the ticket. I got in the car and shrieked away from the curb, gunning to sixty in five, but presumably he'd had enough of me.

When I ran into the hospital, there was a crowd waiting for the elevator. On a weekday morning it can take ten minutes to get to the tenth floor, and after thirty seconds' wait I couldn't take it and ran up the ten flights. I arrived panting and dizzy on 10, immediately bumping into the Howdy Doody doctor from Infectious Diseases, a man Roger and I used to mimic cruelly. He shook his head and said Roger didn't seem able to handle the medication, and I said they had to try it again. I spilled out all the cases the underground had armed me with. An intern heard this whole exchange, watching it like a tennis match. Howdy shrugged and appeared to wash his hands of the matter, but the intern decided to be on *my* side as I stormed away to Roger's room.

The first thing that unnerved me when I got inside was seeing his penis hooked up to a catheter, always a wincing business for a man to witness. The nurse said Rog was comfortable and that he'd been communicative, but for the present he appeared too sick to notice I was there. In any case, I was madly scrambling around making calls to Cope and ordering the intern to speed up the drug from the pharmacy. Finally the nurse realized who I was, and she was my right hand from then on. I'd been there a few minutes, setting up command, when Roger began to moan. It was the saddest, hollowest sound I've ever heard, and loud, like the trumpet note of a wounded animal. It had no shape to it, nothing like a word, and he repeated it over and over, every few seconds. "Why is he doing that?" I asked the nurse, but she didn't know. I assumed he must be roaring with misery and anxiety, and he hadn't had any Xanax since the previous day. I ordered a tranquilizer and told him everything I was doing. It wasn't till ten weeks later, on New Year's Day, that I understood the trumpet sound. I was crying up at the grave, and started to mimic his moaning, and suddenly understood that what he was doing was calling my name. Nothing in my life or the death to come hurts as much as that, him calling me without a voice through a wall he could not pierce.

Within fifteen minutes the intern came in with a shot to relax him, and right after that they began the ampho drip. I was on the phone to Jaimee constantly, the two of us gnawing our hearts as we waited to see if he'd have convulsions. Meanwhile the nurse taught me to communicate with Rog by telling him to blink when I asked him a question. *Can you hear me, Rog*? And his eyelids fluttered. It was such a stunning gift to have him back, tapping through the wall like that. Thirty or forty times in the next hour I made him do it again, lobbing him yes questions and cheering at the reassuring flutter of his eyelids. I kept telling him how much of the drug still had to go in. I talked and talked, excitedly declaring that we were home free. It was working. We were going to bring him back. I held the phone so Jaimee could talk in his ear, and he blinked to say he heard her.

I don't regret a syllable of our manic cheer. I wouldn't have wanted the last he heard from me to be moaning and grief. We were pulling through, as we always did. I asked a friend with a thousand nights' experience of young men dying, How much pain was Roger in that last twenty-four hours? I've heard all the tales of the tribe now about the pounding headaches of crypto. He said the harder thing for Roger than the pain would surely have been the consciousness of his final imprisonment and exile from me. I know he's right because it comes to me in nightmares over and over, the last claustrophobia, no way to touch your friend again or say good-bye as you spiral down. At least we had that queer and eloquent hour of the eyelids, and then he fell asleep.

He took the whole dose of the ampho without a convulsion. After an hour and a half of ceaseless monitoring, the nurse and I actually began to relax, proud of ourselves and how we had shown Dr. Howdy what for. She told me her son was gay, in his mid-twenties, and she considered it her duty to work lovingly with AIDS patients, "so maybe if someone ever has to take care of him, they'll treat him like a son." I was telling her about Roger, what a wise and giving man he was, about our life together—when suddenly out of nowhere he began to breathe strangely. Deep heaving breaths, expelled with explosive release. I could see the nurse's face go pale. "What is it?" I asked.

"Not good," she replied fearfully, running out of the room. The curious helpless breathing continued, like a storm inside him, while I sat there utterly still. Then six or eight different people rushed in, all the interns and the nurses off the floor. They stared at him and jabbered at each other in their own terrible shorthand. Finally one of them turned to me. "Is there a living will? What do you want us to do?"

Nothing. Because that is the point of the living will he'd signed, that we couldn't take him to intensive care and put a tube down his throat. The breathing had leveled out, but his temp was shooting up, and they expected heart failure at any moment. What had happened was that the meningitis had crept to a certain watermark in his brain, and the terrible breathing—Cheyne-Stoking, it's called—was the start of the final drowning. They paged Cope, and he ordered something to reduce the swelling in the brain. Gradually the temp came down to normal, and within the hour he was sleeping deep and easy. He looked most vividly well, in fact, his weight normal and his color good. They gave him oxygen and shouted his name and lifted the side of his head from the pillow, but it slumped back without any muscular life. The battle was over.

I walked through the rest of it numb and lost, borne along by the new and ghastly rituals of separation. Yet I was curiously abstracted too, and unable to cry. The fight had gone out of me, there being no point anymore. Alfred came over and stood outside the room for eight hours, making whatever calls needed to be made. A half-dozen friends came streaming in, and I talked to them all from inside a bell jar. I didn't want to talk to anyone long, because I had to keep going over to Rog to kiss him and tell him I loved him. Everyone always said hearing was the last to go, and I didn't want him to miss a syllable of me before he left, even if he only heard it in a deep and thoughtless dream.

His mother called from Denver to say their flight had been delayed and they were running to catch another. "He's not going to make it this time," I told her. None of us thought he could possibly last till they arrived. Joe Perloff came and sat with me for a while, propping me up with talk of Roger's courage. Dennis Cope, who had fought with us in the trenches for nineteen months, came in and stayed the longest. "What am I going to do without him?" I asked in a hollow voice, and Cope replied immediately, with great force and conviction. "Write about him, Paul," he said. "That's what you have to do."

Sheldon came by but couldn't bring himself to step into room 1010. I had to go to him in the lounge, where he said he didn't know what to say. Cope returned at ten and waited till Al and Bernice arrived. When they walked in they greeted him warmly, not looking toward Rog right away, thanking the doctor for all the long fight. Al gripped my shoulder and declared, his voice breaking, "This boy took care of him like a mother." Then Bernice went to the side of the bed, touched Roger's hand and said, "Good night, sweet prince." But they held their tears, those two, because they had sworn since the very beginning of the end to be strong for me.

After an hour or so they left, and then I dispatched each of the friends. Though Rog had been expected to die by seven or eight, Cope told me at eleven that if he was still alive in the morning they would give him another dose of the ampho. It was all unpredictable now, and Roger might even resurface again, one way or another. Or maybe he would go on like this for hours or even days.

Finally it was Rog and me alone, late at night in the quiet, the way it had been all summer. Still I would not cry, because I wouldn't let him hear sorrow. I spent all my own endearments—*my little friend*—and sat till four o'clock. When I'd kiss his forehead I could still smell the freshness of the shampoo. I called Sam at four and said I was ready to leave, and we talked awhile about whether I needed to be there for the actual moment. I didn't, I don't know why. I clipped a lock of his hair, which got lost in the chaos of the following day. I slipped off his father's sapphire ring, which the nurse had taped to his finger. I said what half good-bye I could. *You're the best*, I whispered as I walked out the door, what I always said when I left his room at night.

I drove home trying to beat the dawn and knew it would not even start until morning. Waking teaches you pain. The parents were in the front bedroom, so I took a Dalmane and curled up in Roger's bed, where I still sleep every night because he is nearer there than anywhere else in the house. When the phone rang at six I drifted out of bed and went into the darkened study. Bernice was standing in the hallway door, and we held each other as the machine answered the phone. After the beep, a voice said: "This is UCLA Medical Center calling. Mr. Roger Horwitz died at 5:42 A.M. this morning, October twenty-second." Bernice and I hugged each other briefly, without a word, and I swam back to bed for the end of the night, trying to stay under the Dalmane. Putting off as long as I could the desolate waking to life alone—this calamity that is all mine, that will not end till I do.

CPSIA information can be obtained at www.ICGtesting.com
Printed in the USA
LVOW131950220413

330356LV00001B/1/P